Thatijanne Santos G. de Carvalho
Carlos Eduardo do Prado Saad
Márcio Gilberto Zangeronimo

Management of Calopsitas (Nymphicus hollandicus) kept in captivity

Thatijanne Santos G. de Carvalho
Carlos Eduardo do Prado Saad
Márcio Gilberto Zangeronimo

Management of Calopsitas (Nymphicus hollandicus) kept in captivity

Behaviour of cockatiels (Nymphicus hollandicus) in captivity

ScienciaScripts

Imprint
Any brand names and product names mentioned in this book are subject to trademark, brand or patent protection and are trademarks or registered trademarks of their respective holders. The use of brand names, product names, common names, trade names, product descriptions etc. even without a particular marking in this work is in no way to be construed to mean that such names may be regarded as unrestricted in respect of trademark and brand protection legislation and could thus be used by anyone.

Cover image: www.ingimage.com

This book is a translation from the original published under ISBN 978-613-9-65367-6.

Publisher:
Sciencia Scripts
is a trademark of
Dodo Books Indian Ocean Ltd. and OmniScriptum S.R.L publishing group

120 High Road, East Finchley, London, N2 9ED, United Kingdom
Str. Armeneasca 28/1, office 1, Chisinau MD-2012, Republic of Moldova, Europe
Printed at: see last page
ISBN: 978-620-7-87236-7

Summary

CHAPTER 1

Introduction

The cockatoo *(Nymphicus hollandicus)* is an Australian bird, belonging to the Kingdom *Animalia*, Phylum *Chordata,* Class *Aves,* Order *Psittaciformis* and Family *Cacatuinae* (BIRDLIFE INTERNATIONAL, 2012), representing the smallest cockatoo in the world.

The animals belonging to the *Psittaciform* Order, although they vary greatly in size, colour and weight, have very striking characteristics that make them easy to recognise immediately, such as a short, high, curved beak with a wide base, rounded, well-moving maxilla articulated to the skull and with extension movements that increase the power of the beak, used to break resistant seeds. They have a thick, sensitive tongue rich in taste buds (COLLAR, 1997). Their feet are zygodactyl, i.e. they have two toes facing forwards and two backwards (SICK, 1997).

Cockatoos are small compared to cockatiels, measuring around 30 cm in length (head-tail) and weighing approximately 1,000g live weight (FORSHAW, 1989). They have an erect crest on the top of their head made up of differentiated feathers, called a tuft. As well as being extremely attractive and showy, they can live up to 25 years of age (TORLONI, 1991). They are nomadic birds and live in flocks, the average size of which is 27 birds (JONES, 1987).

The cockatiel was described by the Scotsman Robert Kerr in 1792 as *Hollandicus psittacus* and was later transferred to its own genus, *Nymphicus,* by Wagler in 1832. It was first described in books in 1793, but became popular worldwide in the 19th century, when the English ornithologist and taxidermist John Gould wrote his book *Birds of Australia*. By 1864 it had become well known to the British and by 1884 it was well established in European aviaries. However, the massive spread of this bird only occurred with the appearance of the first colour mutation, the harlequin, shortly before 1950. From then on, other colour patterns were established, gaining great popularity, practically equalling that of the Australian parakeet *(Melopsittacus undulatus)* (AUSTRALIAN MUSEUM, 2006).

In Brazil, the species began to be introduced in the 1970s (MATHIAS, 2008) and, according to Annex 1 of Ordinance No. 93 of 7 July 1998 of the Brazilian Institute of the Environment and Renewable Natural Resources (IBAMA), the species *Nymphicus hollandicus,* including its mutations, is considered to belong to the domestic fauna throughout the national territory (BRASIL, 1998), even though it is an exotic bird of Australian origin.

Every year, the breeding of cockatiels in Brazil grows, becoming one of the best-selling psittaciformes, with the vast majority of specimens found in private commercial breeding centres and in commercial establishments (BENEZ, 2001). However, information on the species is scarce in the literature, which limits its breeding in captivity.

CHAPTER 2

Birds and homeothermia

Animals have various functional systems, which control body temperature, nutritional status, social interactions and others (GUYTON & HALL, 2002; BROOM, 1981). Together, these functional systems allow the individual to control their interactions with their environment and thus maintain each aspect of their state within a tolerable range (BROOM & JOHNSON, 1993).

Homeothermic animals maintain their body temperature within certain relatively narrow limits, even if the ambient temperature fluctuates and their activity varies intensely (BRIDI, 2006). Birds, being homeothermic animals, have a thermoregulatory centre located in the hypothalamus, capable of controlling body temperature through physiological mechanisms and behavioural responses, through the production and release of heat, thus determining the maintenance of normal body temperature (MACARI et al., 1994). Around 80% of the energy ingested is used to maintain homeothermy and only 20% is used for production. The core body temperature of birds is around 41.7°C (ABREU & ABREU, 2011).

For each animal, there is a lower or upper temperature limit for thermal comfort (thermoneutrality or comfort zone). There are lower or upper critical temperatures at which, through thermoregulation mechanisms, animals are able to maintain their body temperature. Beyond these ranges, the minimum and maximum critical temperatures are reached, which are the survival limits, so that if these limits are exceeded, the animals succumb (BRIDI, 2006).

Figure 1 shows a simplified diagram of the thermoregulation process (SILVA, 2000), showing the lower critical temperature (LCT) and the upper critical temperature (UCT). Beyond these limits, the animal triggers the thermoregulation process. Up to the hi (lower) and hs (upper) temperature limits, the animal manages to maintain its internal temperature with the mechanisms at its disposal, and when the ambient temperature is lower than hi and higher than hs, the animal manages to survive extreme stress in hypothermia (lower) and hyperthermia (upper), but can no longer maintain a constant body temperature. When the hi and hs limits are exceeded, the animal dies.

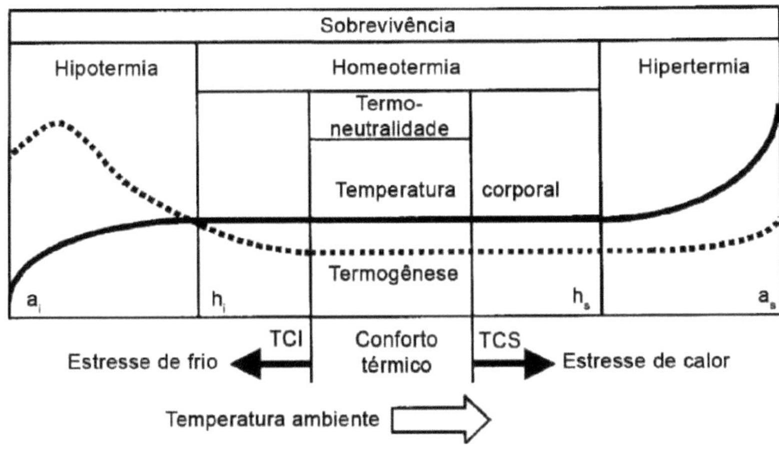

Figure 1 - Simplified diagram of the thermoregulation process in birds. Source: SILVA, 2000.

Pearson (1998) evaluated the development of thermoregulation during the warbling period in cockatiels *(Nymphicus hollandicus)* and hypothesised that, in comparison with other altricial species (which depend on

As a result of parental care) with similar body mass, cockatiel chicks develop homeothermy earlier and consequently have less energy available for growth, resulting in reduced growth rates. They developed endothermic responses in their first week, and were individually effective, with a 75 per cent ability to maintain body temperature during exposure to 20 to 25°C, when compared to adult birds. For cockatiels, the thermal comfort range is between 18.0 and 23.8°C, with temperatures below 12.7°C or above 32.2°C being considered thermal discomfort (GORMAN, 2010).

The effects of temperature and relative humidity on animal thermal comfort are well known, because as these variables increase, the birds' body temperature also increases (MACARI & FURLAN, 2001). As the ambient temperature and/or relative humidity rise above the comfort zone, the birds' ability to dissipate heat decreases. As a result, the bird's body temperature rises and symptoms of heat stress soon appear. The biggest problem in hot and humid tropical areas is excess relative humidity, which makes it impossible for the bird to eliminate heat efficiently through respiration. When the temperature and relative humidity are high, the bird can't breathe fast enough to remove all the heat it needs to dissipate from its body. In this case, the body temperature can rise, causing prostration and death when it reaches 47 °C, which is the bird's maximum vital physiological limit (MITCHELL, 1976, NÃÃS, 1994; RUTZ, 1994). The increase in rectal temperature is also a physiological response to high temperature and humidity conditions that result from the storage of metabolic heat (SILVA et al., 2003). In most birds, the normal cloacal temperature is around 41.7 to 44.4°C (GREENACRE & LUSBY, 2004).

Heat loss is realised in two ways: sensible heat and latent (or insensible) heat. Sensible heat is the heat that causes the environmental temperature around the bird to rise, as a result of factors such as physical activity and increased heat after feeding, occurring through the following processes: radiation, conduction and convection. Birds can lose up to 70 per cent of the extra heat produced through these processes, but the rate of loss is influenced by the environmental temperature (ABREU & ABREU, 2011).

Conduction is the exchange of heat between two touching bodies or even parts of a body at different temperatures. In conductive heat flow, a hot molecule of the body in question collides with a neighbouring cold molecule and transfers part of its kinetic energy to this molecule, and so on, tending towards equilibrium (NÃS, 1989; BAÊTA & SOUZA, 1997). Direct contact is required between the molecules of the bodies involved. Convection occurs when a current of liquid or gaseous fluid, which absorbs thermal energy in a given location, moves to another location, where it mixes with colder portions of that fluid and transfers the thermal energy to them (BRI DI, 2006). Radiation can be defined as the transfer of energy from one body to another through electromagnetic waves (NÃÃS, 1989).

Latent (or insensitive) heat loss occurs through evaporation (a change in the state of water from liquid to gas) and is observed in the skin and respiratory tract. This route is characteristic of species with few sweat glands. It is an exhaustive form of heat loss, since 1 g of water allows the elimination of 530 calories (ABREU & ABREU, 2011).

Thus, when the ambient temperature exceeds the upper critical temperature, one of the first mechanisms triggered is an increase in water intake and a reduction in feed consumption. The increase in water consumption is directly related to the increase in water demand for evaporative and respiratory heat loss (MOURA, 2001). Water consumption for stressed birds doubles that of birds kept at milder temperatures (BONNET et al., 1997). The reduction in food consumption occurs because it reduces the metabolic substrates available for metabolism, thus reducing heat production (BELLAY & TEETER, 1993).

Behavioural adjustments can occur more quickly and with less energy expenditure than many physiological responses, such as a reduction in physical activity, because when birds remain seated and have their wings open, they increase heat dissipation by maximising their body surface area (FURLAN, 2006).

Thus, the compensatory physiological responses of birds when exposed to heat include peripheral vasodilation, resulting in an increase in non-evaporative heat loss, so that birds keep their wings away from their bodies, ruffle their feathers and intensify peripheral circulation (BORGES et al., 2003). Birds can also "open their beaks" in response to high temperatures, i.e. they can increase their respiration rate and the internal area through which air passes in order to increase heat dissipation through evaporation (ABREU & ABREU, 2011). Increasing the respiration rate results in excessive losses of carbon dioxide (CO_2). Thus, the partial pressure of CO_2 (PCO_2) decreases, leading to a drop in the concentration of carbonic acid (H_2CO_3)

and hydrogen (H+). In response, the kidneys increase the excretion of HCO3- and reduce the excretion of H+ in an attempt to maintain the bird's acid-base balance. This change in the acid-base balance is called respiratory alkalosis and can lead to the bird's death (BORGES et.al., 2003).

1.1. environmental indices

Ever since the differences between animals in terms of their ability to cope with the climatic variations of the environment in which they live were recognised, attempts have been made to establish criteria for classifying the various environments and combinations of factors that provide thermal comfort for animals. In this context, various thermal environment indices have been developed, encompassing the combined effect of meteorological elements and the environment in a single parameter (YANAGI JÚNIOR, 2006).

Among the simpler indices, those that involve a smaller number of variables, the temperature and humidity index (ITU) has stood out, as it encompasses only the effects of dry bulb temperature and relative humidity. Examples of the development of these indices are the UTI developed by THOM (1959) and DESHAZER & BECK (1986) (equations 1 and 2, respectively).

$$ITU_{Thom} = t_{bs} + 0,36 \cdot t_{po} + 41,5 \quad (1)$$

$$ITU_{D\&B} = 0,6 \cdot t_{bs} + 0,4 \cdot t_{bu} \quad (2)$$

in which,

tbs = dry bulb temperature (°C)

t_{po} = dew point temperature (°C)

tbu = wet bulb temperature (^{2} C)

For broiler chickens, ITUihom values below 74 indicate thermal comfort for the animal; between 74 and 79 represent situations of alert for producers and danger for production; between 79 and 84 indicate situations of danger with the possibility of production losses and values above 84 indicate emergency situations, requiring urgent measures to avoid the loss of the plant (NSCR, 1976).Originally developed for laying hens, the UTID&B has also been used to quantify broiler production in response to the thermal environment (GATES et al., 1995), 1995), with values above 25 characterising thermal discomfort (GATES et al., 1995; SILVA et. al., 2003).YANAGI JÚNIOR et al. (2001) and TAO & XIN (2003) developed the index of temperature, humidity and air velocity (ITUV) for laying hens and broilers, respectively, subject to the acute condition of thermal discomfort due to heat (equations 3, 4 and 5 respectively).

$$ITUV_{poedeiras} = -15,17 + 18,62 \cdot \left(\frac{t_{bs}}{41}\right) - 0,92 \cdot \left(\frac{DVP_{ar} \cdot \sqrt{v}}{5,8}\right) \quad (3)$$

$$ITUV_{frangos} = \left(0,85 \cdot t_{bs} + 0,15 \cdot t_{bu}\right) \cdot v^{-0,058} \quad (4)$$

$$DVP = \left(1 - \frac{UR}{100}\right) \cdot P_{ws}\left(t_{bs}\right) \quad (5)$$

in which,

tbs = dry bulb temperature (°C)

tbu = wet bulb temperature (°C)

Pws.(tbs) = vapour saturation pressure calculated as a function of tbs, (kPa) v = air velocity (m/s)

The UTI values found for normal thermoneutrality and natural ventilation conditions should be close to 26 (ANGELO et.al., 2014).

Heat exchange by radiation is of great importance in quantifying the thermal environment, especially for facilities with low thermal insulation and animals kept outdoors. In these cases, the use of the black globe temperature and humidity index (BTIH), proposed by BUFFINGTON et al. (1981), is recommended because it incorporates tbs, RH, v and radiation into a single value (equation 6).

$$ITGU = 0,72 \, (t_{gn} + t_{bu}) + 40,6 \quad (6)$$

in which,

tgn = black globe temperature (°C) tbu = wet bulb temperature (°C)

The reference values for thermoneutrality are the same as those suggested for ITUihom: < 74 configure thermal comfort; 74 and 79, alert situation; between 79 and 84, danger situation; > 84, emergency situation (NSCR, 1976).

Another index that depends on the black globe temperature is the thermal radiation load (CTR), which should be as low as possible. The CTR, proposed by Esmay (1969), expresses the total radiation received from all spaces or parts of the neighbourhood (equations 7 and 8 respectively).

$$CTR = \sigma \, (TRM)^{4} \quad (7)$$

$$TRM = 100 \sqrt[4]{2,51\sqrt{v}(Tgn - t_{bs}) + (tgn/100)^{4}} \quad (8)$$

in which,

a = Stefan Boltzmann constant (5.67 x 10^{-8} W m^{-2} K $)^{-4}$

TRM = mean radiant temperature (°K) v = wind speed (m/s) tbs = dry bulb temperature (°K) tgn = black globe temperature (°K)

Although these indices are among the most popular, they have their shortcomings and limitations. Therefore, other indices are constantly being proposed and validated. The major difficulty remains the application of these indices in different environmental conditions (TAKAHASHI et. al., 2009).

CHAPTER 3

Poultry facilities

Buildings represent a significant portion of productive investment and, when not properly planned, can cause serious damage to animal production (HARDOIM, 1998).

The reason for building shelter for animals is to protect them from bad weather. For this protection to be effective and efficient in terms of animal productivity, it is necessary to quantify the interaction between climate, animal and type of shelter (NÃÃS, 1989).

A suitable construction system is able to control the climatic factors that most interfere with thermal comfort inside the building, such as temperature, humidity, solar radiation and wind. In order to obtain a suitable building, it is necessary to take into account, in its planning, the construction materials, the type of animal that will inhabit it and the local climate (BAÊTA & SOUZA 1997; NÃÃS, 1989).

Curtis (1983) defines the environment as the sum of the impact of physical, chemical, biological and social factors that act and interact to influence animal performance. These factors vary according to the season and the location of the region, as well as for reasons intrinsic to the animal itself, such as age, sex and the feed provided. He distinguishes between two classes of environmental modification: primary and secondary. Primary environmental modifications are those related to the enclosure, i.e. those related to the shelter, to the poultry house itself, and which protect the bird during periods when the weather is extremely hot or cold, helping it to increase or reduce its loss of body heat. The primary modifications correspond to natural thermal conditioning. Secondary modifications correspond to the management of the facility's internal microenvironment. They generally involve a higher level of sophistication and include artificial ventilation, heating and cooling processes. Secondary modifications must be carried out after all primary modifications have been exhausted.

1.2. Primary Modifications

The location of the facilities is of the utmost importance for obtaining satisfactory results in the development of the activity. In any case, when setting up new poultry complexes, a prior study or planning of the minimum conditions necessary for the proper functioning of the facilities is always necessary. When planning a site, low lying land should be avoided, so as to avoid problems with high humidity, low air movement and insufficient hygienic sunshine in winter. Attention should also be paid to the possible obstruction of air by other buildings and natural and artificial barriers near poultry sheds, which would hinder natural ventilation and jeopardise thermal comfort in summer. For tropical and subtropical climates, the longitudinal axis of poultry buildings should be orientated in an east-west direction, to avoid overheating due to strong sunlight during the long summer afternoons (TINÔCO, 2001).

The distance between sheds must be sufficient so that one does not act as a barrier to natural ventilation in the other. Thus, a distance of 10 times the height of the building is recommended for the first sheds, and from the second shed onwards the distance should be 20 to 25 times this height. It won't always be possible to meet this recommendation for reasons relating to the area available, topography or work flow, but you should try to get as close to this situation as possible. In the worst case scenario, the distance between sheds should be at least 35 to 40 metres (TINÔCO, 2001).

As for the size of the sheds, according to BAÊTA (1995) quoted by FURTADO et al. (2005), the width of the shed has a great influence on the interior thermal conditioning and its cost, and there is a worldwide trend to design sheds that are 12 metres wide and 125 metres long.

According to AVILA et al. (2007), the ceiling height must be at least 3 metres, and pre-moulded concrete, metal or wooden structures can be used.

Wide eaves prevent the sun from shining into the shed, causing the temperature to rise, and also prevent water from wind-driven rain from entering (SOUZA, 1997). In very rainy regions, a 45° slope to the floor is recommended. In general, eaves of 1.5 to 2.0 metres are recommended on both the north and south sides of the roof, depending on the ceiling height and altitude (TINÔCO, 2001).

The use of lining is crucial for good ventilation performance in poultry houses, as it reduces heat entering the facility in summer and heat escaping in winter (SOUZA, 2005). According to Costa (1982), this reduction is 62 per cent when moving from an unlined shelter to a shelter with a simple unventilated 6 mm duratex lining and 90 per cent in the case of a ventilated lining.

The roof is one of the main factors influencing the incident radiant heat load and acts on the internal environment as a result of the roofing material, reducing the heat flow inside. A good roofing material should have high solar reflectivity combined with low thermal emissivity and absorptivity. Although clay tiles have a better thermal performance, roofing made with corrugated asbestos cement tiles has a lower construction cost than that made with clay tiles, mainly because the supporting structure is lighter and less labour is required (TCPO 7, 1980), it is also quicker to make and easier to clean, which justifies the preference for this type of roofing by poultry farmers. The best material for reducing the radiation load is clay tiles, followed by asbestos cement tiles painted white and aluminium, respectively (BRIDI, 2006).

The slope of the roof affects the thermal environment inside the shed, so the steeper the slope, the greater the natural ventilation. Inclinations between 20 and 30° have been considered adequate, taking into account structural and thermal environmental conditions (TINÔCO, 2001).

For sheds with widths of 8.0 metres or more, the use of a rooflight is essential. Its function is to allow hot air to escape, especially during the hot season (TINÔCO, 2001). It is recommended that the rooflight be built in two sections, arranged lengthways along the entire length of the roof, equipped with a system that allows it to be closed easily and with wire mesh to prevent birds from entering. It must allow for an opening of at least 10 per cent of the width of the aviary, with roofs overlapping at a distance of 5 per cent of the width of the aviary or at least 40 cm. The ends of the roof lantern should be a maximum of 5 cm from the roof opening to prevent rain from entering the poultry house (ABREU & ABREU, 2000a).

The ideal type of shade, according to Kelly et al. (1950), although its comparative values are difficult to measure, is the shade produced by trees, because vegetation transforms part of the solar energy through photosynthesis into latent chemical energy, reducing the effects of insolation during the day. Thus, the use of tall trees can produce a mild microclimate in the facilities. SILVA & NÃÃS (1998), studying the influence of afforestation on the thermal performance of poultry houses, concluded that afforestation reduced the internal temperature of poultry houses by approximately 10.3%. Unit egg production was 23.1% higher in the wooded area than in the non-wooded area. It is worth noting, however, that even in the shade, the bird is subject to indirect radiation from the distant sky, the shaded ground, the heated ground and structures that are close to the site (TINÔCO, 2001).

Windbreaks are natural or artificial devices designed to stop or at least reduce the action of strong winds on crops and buildings. They can also be defined as structures perpendicular to the prevailing winds, whose functions are to slow them down and reduce the damage they cause. Most of them are natural, made up of clumps of vegetation (BAÊTA & SOUZA, 1997).

From the point of view of animal environment engineering, a very effective resource in the air conditioning of poultry houses is the increase in natural ventilation, which can be achieved through the appropriate architectural design of the houses (TINÔCO, 2001). Natural ventilation is the normal movement of air that can occur due to differences in pressure caused by the action of the wind (dynamic ventilation) or temperature (thermal ventilation) between two media (ABREU & ABREU, 2000b). By properly controlling the entry of heat into the poultry house and facilitating the exit of the heat produced, ventilation becomes a complement to comfort requirements.

Adequate ventilation is also necessary to eliminate excess humidity from the environment and the litter, which comes from the water released by the birds' breathing and through the droppings, and to allow air renewal by regulating the level of oxygen needed by the birds, eliminating carbon dioxide and fermentation gases (ABREU and ABREU, 2000b).

According to Souza (2005), in laterally open poultry houses, the management of curtains is fundamental to obtaining a healthy flock, high welfare and productivity throughout the flock's growth period. NÃÃS (1989) suggests that there should be a difference in level between the air inlet and outlet openings, and that they should be located on opposite walls so that ventilation is efficient. Obstacles inside the building or any protrusion on the façade alter the direction of the air flow.

Artificial ventilation is produced by special equipment such as exhaust fans and ventilators. It is used whenever natural ventilation conditions do not provide adequate air movement or temperature reduction. It has the advantage of allowing filtering, uniform and sufficient air distribution in the poultry house and being independent of atmospheric conditions (ABREU & ABREU, 2000b).

There are two ways of artificially promoting air movement: a negative pressure system (exhaust) and a positive pressure system (pressurisation). Positive pressure equipment (fans) is installed to push the outside air into the facility, forcing the inside air out. The fans should be placed at mid-height, slightly inclined towards the floor, but not directly above the birds. The negative pressure system is realised by means of exhaust fans, and is most often a very efficient system. In this process, the air is forced through the exhaust fans from the inside out, creating a vacuum inside the facility. The system creates a difference in air pressure between inside and outside and the air escapes through openings (DAMASCENO et.al., 2010).

Among the secondary modifications, the adiabatic evaporative air cooling system has expanded rapidly in places affected by heat stress. This system is simple, practical and has a good cost/benefit ratio, which has pleased many producers (ARCARO JÚNIOR et al., 2003). The principles of its functionality consist of changing the psychrometric state point of the air, towards higher humidity and lower temperature, through the contact of the air with a humidified or liquid surface (SMITH et al., 2006). The adiabatic evaporative cooling system, as it is used to air-condition facilities, consists of forcing external air through a moistened porous material, using fans or exhaust fans (VIGODERIS, 2002).

According to Abreu et. al. (1999), in Brazil, the evaporative cooling systems commonly used inside poultry houses basically consist of *padcooling* and nebulisation systems (low and high pressure). Some producers also use a sprinkler system on the poultry house roof to minimise the effect of the radiant heat load on the birds.

Pad cooling systems require mechanical ventilation to force air through the evaporative panels and are commonly used in air-conditioned poultry houses consisting of a fully automated system with negative ventilation in a wind tunnel. The evaporative panels used in this process are usually made of special cellulose material, kept constantly moist, through which the air passes and cools down before entering the interior of the poultry house. The water supply in the evaporative panels can be carried out by water pipes installed at the top or by sprinkling water in front of the evaporative panel.

The misting system is made up of misting nozzles that break up the water into tiny droplets and distribute it inside the poultry house in the form of a jet of water. This system can be operated at high or low pressure. The higher the system's working pressure, the greater the break-up of the water droplet.

According to Abreu et al. (2000b), there are difficulties in maintaining the ideal comfort temperature for chicks. In the early stages of life, chicks do not yet have a developed thermoregulatory system, and it is necessary for the ambient temperature to be in suitable ranges according to their development. The most commonly used equipment for heating chicks are gas or electric hoods, infrared lamps considered to be electric heaters and furnaces that use wood as a heating source.Lighting management can also influence the productive and reproductive performance of certain animal species. According to BAÊTA & SOUZA (1997), poultry and horses are long-day breeders, while sheep and goats are short-day breeders. For cattle and pigs, there is no influence of photoperiod on reproduction processes. In the poultry industry, light management has been successfully applied to increase the quantity of eggs produced and the production of heavy birds.

1.4. Facilities for psittacines

The Brazilian Institute for the Environment and Renewable Natural Resources (IBAMA), through ordinances 117 and 118 of 15 October 1997, which regulate the breeding of native wild species and the sale of live, slaughtered animals, parts and products of native fauna, has led to an increase in the legal breeding and sale of these animals in recent years. Currently, it is possible for pet shops registered with IBAMA to sell psittaciforms bred in breeding centres duly registered with the institute. These ordinances are aimed at collaborating with the conservation of our fauna and combating the illegal trade in wild animals, as it is believed that the illegal trade should progressively decrease as there is the possibility of acquiring animals in a legal and reliable manner, with correct documentation, controlled health and origin and adapted to captivity (ALLGAYER & CZIULI, 2007).

However, there are difficulties when it comes to breeding wild animals in captivity, not least the high level of bureaucracy involved in becoming a commercial breeder authorised by IBAMA. There are a number of legal requirements that must be met, which discourages those interested in breeding. In addition, there is little literature on the commercial breeding of wild birds, which means that breeders are forced to work without a proper scientific basis, i.e. many procedures are learnt by observation while the business is developing (ANTONIALLI et. al., 2004).

Allgayer & Cziuli (2007) advocate that the quarantine should be away from the other birds' ponds (exhibition or breeding), and should have equipment (brooms, buckets, feeders, drinkers) to avoid the possible transmission of diseases to the poultry. As soon as they arrive in quarantine, the birds must be identified and marked *(ring/microchip)*. A clinical examination must be carried out and biological samples taken. During the quarantine period, it is essential to observe the birds on a daily basis, as during this phase they are adapting to the nutrition recommended at the breeding centre. All clinical signs compatible with an illness should be investigated. During this quarantine period, prophylaxis is also carried out, which must be established for each breeding centre. Once the quarantine period is over and the birds are fit, they can be transferred to the permanent flock.

Display enclosures can be mixed or individual, always obeying the occupancy density for each species. They can be of different sizes and represent different ecosystems. It is important for these enclosures to offer several nests, perches and feeding areas, as well as escape areas for the birds.

In the case of psittaciformes, there are several types of vivarium. They can be conventional (on the ground) or suspended cages (more commonly used by breeders). It's important that the cages are separated from each other so that the birds don't become aggressive towards their neighbours and end up suffering trauma when they come into contact through the screens. The best reproductive results are undoubtedly obtained in commercial installations with metal cages suspended from mesh. These cages can be inserted into open sheds, with the nest and feeder located inside. The size of the cages can vary according to the species used.

The food storage area is essential and must be ventilated to prevent damp and protected from rodents.

According to Perencin et. al. (2011), facilities for psittacines should follow these recommendations:

a) Protected from wind and weather;

b) Provide refuge from possible predators;

c) Being in high places;

d) Keep cages spacious, allowing for flight and ample movement;

e) Preferably suspended vivariums, avoiding contact between the bird and the ground or the place where faeces and food are deposited;

f) The perches should be rough but not abrasive, with different diameters and at different heights;

g) The positioning of the perches must allow the animal to move fully, avoiding contact with the bars in order to prevent any kind of accident;

h) Feeders and drinkers should be made of rigid, non-toxic, washable material;

i) The enclosure should be enriched with toys and objects, which is important to avoid any development of behavioural disorders.

j) It's a good idea for the cage to have a removable tray to make it easier to change the absorbent paper at the bottom, which should be done daily.

Many people don't know that mistreating animals is a criminal offence under federal law. Even those who do know think that mistreatment can be summarised as aggression, but it actually goes much further than that. In addition to aggression, it is also considered mistreatment not to feed them properly, to leave them locked up in a place unsuitable for their size, to leave them exposed to the sun and rain, not to take them to the vet when necessary, among many other examples (PERENCIN et. al., 2011).

Therefore, animals bred in captivity must be monitored in order to control the various production factors, such as food, health, environment and others, with a view to their reproduction (TRUTH, 2004). In addition, all the material used in the enclosures must be cleaned and sanitised constantly, thus providing a suitable environment for the animal, with good sanitary conditions and a beautiful appearance of the enclosure where the animal resides. The materials used to promote animal welfare, such as trunks, perches, leaves and sand should be changed whenever necessary (VIEIRA et al, 2008).

CHAPTER 4

Social behaviour of cockatiels in captivity

Humans have always had a great interest in psittaciformes, making them pets (FORSHAW, 2010), due to the group's inherent characteristics, such as the ability to imitate sounds and high levels of learning (GODOY, 2006; SICK, 1997). The cockatiel is one of the most popular bird species as a companion animal in the United States. Due to their large numbers, endurance and ease of reproduction in captivity, they have been considered an ideal model species for studying the social behaviour of psittaciformes (YAMAMOTO et al., 1989).

The study of the social behaviour of psittaciformes has several important applications. Information on aggression, dominance relationships, affiliative and reproductive behaviour can help to improve the conditions of birds kept in captivity by providing suitable environmental conditions and addressing both the physical, social and individual needs of the birds (COLLAR & JUNIPER, 1992).

The importance and significance of dominance interactions in psittacine social groups was evaluated by Seibert (2003). Stability of social groups requires both mutual recognition of members and a system of resource allocation limits. Once relationships are established, there is no consistency in social interactions, resulting in fewer, or less intense, aggressive assertions of dominance (BERNSTEIN, 1981). These relationships have the function of reducing the occurrence of competitive conflicts between members of a social group.

Agonistic behaviour consists of two actions, aggressive and submissive, within the context of a social interaction (WILSON, 1975). Agonistic encounters are most often observed when relationships are unclear, such as the introduction of new individuals. Dominance relationships also seem to require periodic reinforcement, even in the absence of a stimulus, to avoid extinction. Subordinate individuals respond to aggressive behaviour carried out by individuals with appeasement or submissive signals to the hierarchically superior individual. Submissive postures allow them to avoid combat (BERNSTEIN, 1981).

The advantages of the higher flock *status* that birds occupy have not been determined for most psittacine species. Hierarchically superior individuals may have greater access to feeding or roosting sites, less visibility to predators or more mating opportunities. Aggressive encounters in a group of orange-fronted parrots *(Eupsittula canicularis)* were most frequent during feeding, followed by bathing or searching for places to roost (HARDY, 1965). However, no aggression occurred in the context of foraging in a captive flock of cockatiels *(Nymphicus hollandicus),* but higher-ranking males seem to have greater access to mates and preferred nest boxes (SEIBERT & CROWELL-DAVIS, 2001).

In more widely studied bird species, males show a higher frequency of aggressive behaviour than females (WOOLFENDEN & FITZPATRICK, 1977; WINGFIELD et. al., 1987; JACKSON, 1991; WILSON, 1992; NOL et. al., 1996; SEIBERT & CROWELL-DAVIS, 2001). Seibert & Crowell-Davis (2001), studying a captive flock of cockatiels, found that females were significantly more likely to aggress against other females than against the males in the flock. There was no significant difference for males and their adversaries. Female competition for access to mates has been suggested as an explanation for these gender differences in female aggression (SANDELL & SMITH, 1997; TARVIN & WOOLFENDEN, 1997).

The behaviours of the different species of psittaciformes are similar, so, taking into account the observations of cockatiels in captivity and the results described by Prestes (2000), who evaluated the shark parrot *(Amazonapretrei)* also in captivity, we can classify the social behaviours of cockatiels into two categories: social non-agonistic and social-agonistic.

4.1. *Non-agonistic social*

a) Social grouping - Approach

Walking or flying, one bird approaches the other. If it is not accepted, it immediately performs another behaviour. It may also approach another bird to take food from it (Figure 2).

Figure 2 - Representation of grouping-approach behaviour in psittacines, from the non-agonistic Social category. Source: PRESTES, 2000.

b) Social grouping-Attendance

One bird moves away from the other for no apparent reason, or because it doesn't accept their proximity. Generally, they move away laterally (Figure 3).

Figure 3 - Representation of grouping-attachment behaviour in psittacines, from the non-agonistic Social category. Source: PRESTES, 2000.

c) Social cleaning

One bird approaches the other. If it is receptive, it strokes the other individual's feathers with its beak, usually on the apex of the head, throat, neck, both the anterior and posterior lower parts and the upper neck, rarely on the primary rémiges. If the recipient accepts this cleaning well, they close their eyes and turn their head to the side of their mate. On other occasions, it has been observed that the approaching bird tilts its head towards the other, ruffling the feathers on its head (Figure 4).

Figure 4 - Representation of cleaning behaviour in psittacines, from the non-agonistic Social category. Source: PRESTES, 2000.

d) Request food

One bird approaches the other and, moving its head up and down, emits a "crãck, crãck, craãck..." sound. It insists several times. It keeps its body in a lower position than the other. It lowers its head and stays there for a few seconds. The tail is lower than the body (Figure 5). This behaviour is exhibited by the chicks, who often ask their parents for food.

Figure 5 - Representation of food-seeking behaviour in psittaciformes, from the non-agonistic Social category. Source: PRESTES, 2000.

e) Passing on food

The bird approaches another individual. It makes greater and faster efforts, regurgitating and passing the food through its beak to the individual next to it. The receiving bird may or may not have asked for it. After a short time, it repeats this behaviour. The bird receiving the food tilts its head back approximately 45° from the horizontal position, and the bird passing the food forces its head downwards four or five times, regurgitating the food, remaining in beak-to-beak position and emitting a "trãáã, trãáã, trãáã" sound each time the food is passed from one individual to another (Figure 6).

Figure 6 -   Representation of the non-agonistic social category   of   food-grabbing   behaviour   in psittaciformes. Source: PRESTES, 2000.

It can also happen that the birds share the same food (Figure 7). In a study by Andrade & Azevedo (2013), it was observed that it was common for true parrots *(Amazona aestiva) to* share the same food enrichment, and as soon as they were introduced into the enclosure, the birds moved towards them and began to explore them.

Figura 7.   Representation of food-sharing behaviour in psittaciformes. Source: PRESTES, 2000.

4.1. *Social-agonistic (bickering)*

One bird chases the other, usually in flight, pecking at it while it is still suspended in the air. It does this several times. When it lands next to you, it pecks you on the wings, back and tail until the aggressed bird surrenders or the aggressive one feels defeated. It can use one of its legs to push away the bird in its path, leaning on its wing. Fights occur because of a dispute over a place to roost or feed, or because one bird approaches another with which it was interacting in a non-agonistic way. When fighting, the wings are slightly raised and pulled away from the body, projecting the head forwards, pecking at the desired area of the other bird's body. The aggressor bird is usually on a higher plane and rests one leg on the landing substrate and the other on the aggressed bird. The part of the body that is hit can be as varied as possible, but it is usually the legs and wings. The other bird defends itself with its beak. Both make sounds (Figure 8). Male cockatiels are generally more aggressive than female cockatiels (SEIBERT & CROWELL-DAVIS, 2001).

Figure 8. Representation of fighting behaviour in psittacines, from the Social-agonistic category. Source: PRESTES, 2000.

CHAPTER 5

Physiology of digestion

5.1. Eating habits

Among birds, psittacines are the most selective when it comes to food, as they have a thick tongue rich in taste buds, giving them a well-developed palate (CUBAS et al., 2006). Although research into the olfactory capacity of psittaciformes is scarce, several species of birds use olfactory cues to locate food, as well as for orientation and navigation to return to the nest, reproduction, paternity and the selection of materials for nest construction (GRAHAM etal., 2006).

The cockatiel is classified as a granivorous bird (KOUTSOS et al., 2001a) and its diet in natural conditions consists of a wide variety of seeds, fruits, leaves, flowers and insects (HARCOURT-BROWN, 2003), selecting its diet in arid regions of Australia (FORD, 1974). In the wild, cockatiels feed mainly on seeds, with sorghum seed being the most commonly eaten. They are nomadic birds and, unlike other psittaciformes, which feed at the top of trees, they have the habit of feeding on the ground, in flocks, the average size of which is *IT* birds (JONES, 1987).

In captivity, the nutrition of psittaciformes in general has basically two possibilities: the provision of seed mixtures and/or pelletised feed as the basis of the diet. The availability of commercial feed for ornamental birds is still very small when compared to potential consumption. In addition to providing the nutritional principles in adequate quantities to meet the birds' needs, commercial feeds must include a series of other nutritional aspects such as raw material quality and palatability (MACHADO & SAAD, 2000). Due to the limited availability of specific commercial feed for psittaciformes, it is common practice in breeding centres and zoos to offer non-specific feed, such as broiler feed or dog food, as well as a mixture of various seeds, such as birdseed, peanuts and sunflower seeds, which are offered at will (SAAD et.al, 2007).

When a mixture of seeds is offered in a diet for psittaciformes, the animals prefer the lipid-rich seeds over the carbohydrate-rich seeds. This behaviour is not the result of the energy content of the seeds, but of selectivity related to the palatability of the ingredients (LORO PARQUE, 2009).

Koutsos et al., 2001a states that for a male cockatiel the percentage of 11% protein (soya bean-based diet) supplemented with methionine (0.51% methionine and 0.77% lysine) was the best ratio for adequate maintenance and preventing obesity. It also suggests that cockatiels are able to regulate the catabolism of amino acids in diets with high levels of protein, similar to what happens in omnivorous animals. Furthermore, while in commercial poultry farming, chickens need 3.3 per cent calcium in their diet for adequate calcification, cockatiels can lay eggs with normal calcification with 0.35 per cent calcium in their diet.

Koutsos et al., 2001b found that cockatiels are more likely to have problems with diets containing excess vitamin A than with vitamin A deficiency. The cockatiels survived for two years on diets containing 0 pg of vitamin A, with no changes in their plasma retinol. This shows that these animals have adapted to the nutrient-poor diets they consume in the wild. It is suggested that the vitamin A requirement for maintenance is less than 2000 UI/kg, but this has not been determined.

5.2   Digestive system

The anatomy of birds is different from that of other animal species. Among the different systems, the digestive system stands out. Each bird's digestive tract is adapted to process and utilise the food available in its *habitat as* efficiently as possible (DUKE, 1996; POUGH et.al., 1999). This evolutionary capacity has allowed the survival of today's diverse bird populations, which have created their own ecological niches according to the food resources available (PETERSON, 1971).

The anatomy of the digestive system of psittaciformes differs from that of other bird species. *Psittaciformes are* characterised by the absence of cecum (or functional vestigial cecum), the presence of a short colon and a high rate of food passage through the digestive tract (RITCHIE et al., 1994). However, the digestive tract itself shows little variation between psittacine species (SANTOS et al., 2012).

According to Bennett & Deem (1996), the digestive system of *Psittaciformes* (Figure 9) is made up of the beak, oesophagus, stomachs (proventriculus and ventriculus), small intestine and large intestine.

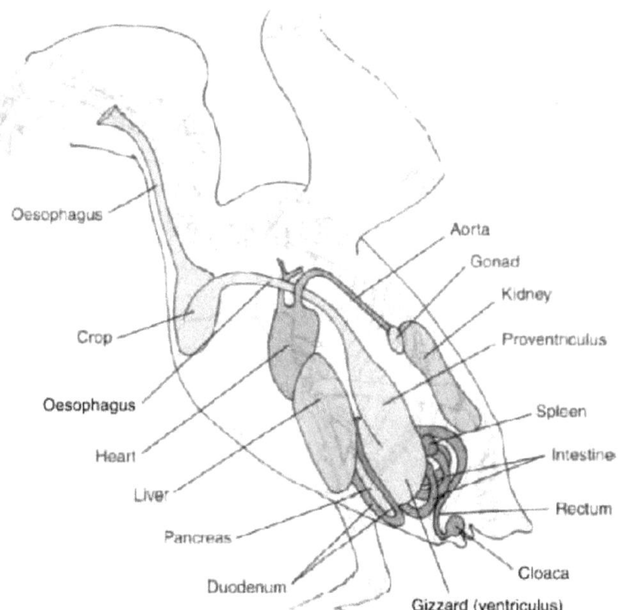

Figura 9.   Psittacine digestive system. Source: HARCOURT-BROWN & CHITTY, 2005.

The process of digestion in birds begins by grasping the food, the lips and teeth of which are replaced by a horny formation: the beak. Birds' beaks grow continuously and are a dynamic structure made up of bone, vascular layers, keratin, dermis, joints and a germ layer. The keratinised sheath covering the upper and lower beak is called ranfoteca (Figure 10) and can be divided into rhinotheca (maxillary keratin) and gnatotheca (mandibular keratin).

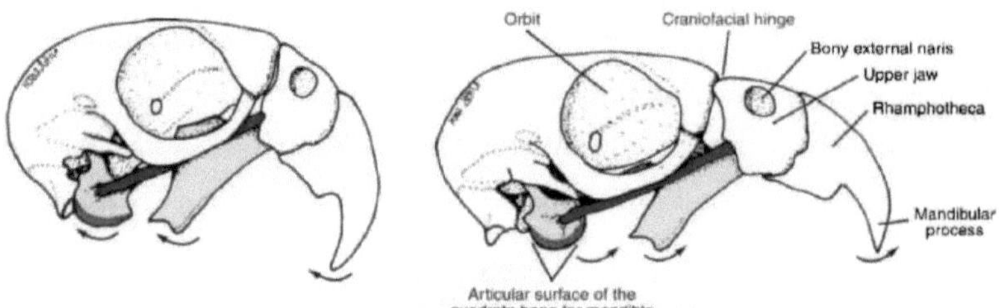

Figura 10.   Lateral and ventral views of the skull of a blue-headed parrot *(Pionus menstruus).* Source: HARCOURT-BROWN & CHITTY, 2005.

Birds have neither a soft palate nor a clear constriction separating the mouth from the pharynx. Therefore, oropharynx indicates the combined cavity that extends from the beak to the oesophagus (DYCE et. al., 1997). The walls of the oropharynx contain numerous mucus-secreting salivary glands. Swallowing is purely mechanical, with the food bolus being swallowed by sudden movements of the head (RITCHIE et.al, 1994).

The oesophagus is divided into the cervical oesophagus, the ingluvial oesophagus and the thoracic oesophagus. Below the oesophageal epithelium are salivary mucus-producing glands that resemble salivary glands (SCHMIDT et. al., 2003). Among birds, the only species whose saliva contains amylase is the goose. The digestive role of saliva is weak, favouring only food transit (RITCHIE et.al, 1994).

The ingluvium is a saccular structure that forms part of the oesophagus, located at the transition between the cervical and thoracic portions of this organ, which has the capacity to dilate to become highly distensible (SCHMIDT et. al., 2003). It is responsible for storing food for short periods of time, specifically when the muscular stomach (ventricle) is full. It is in the ingluvium that fermentation and soaking of food with mucus occurs, preparing it for subsequent gastric digestion (McLELLAND, 1986; DYCE et. al., 1997). When fully distended, the ingluvium is almost transparent and varies in size and shape according to bird species. The crop serves as a storage organ that partially regulates the entry of ingested food into the gizzard. It allows the bird to consume a large amount of feed in a short period of time and digest it later. Considerable microbial growth occurs in the crop, which can contribute to the digestion of food (CHAMP et al., 1983).

In psittaciforms, the proventriculus is a thick organ with an elongated, fusiform appearance, directed ventrally to the left and situated on the left antimer of the floor of the thoracoabdominal cavity (McLELLAND, 1986). Its wall is predominantly made up of tubular glands. These glands contain a single layer of cells that produce both pepsinogen and hydrochloric acid (SCHMIDT et.al., 2003).

The ventricle (gizzard) is a hollow, rounded organ (SCHWARZE, 1980), made up of thick external musculature, necessary for crushing ingested food (SCHMIDT et. al., 2003). It can be said that it functionally compensates for the absence of teeth by grinding pre-digested food (POUGH et. al., 1999). It has a highly developed circular musculature, whose rhythmic and strong contractions are responsible for crushing the ingested food (MACARI et al., 1994).

The junction between the proventriculus and the ventriculus is the isthmus. It is characterised by a very short junction where there is a transition between the proventricular and ventricular glands (SCHMIDT et. al., 2003).

The small intestine is anatomically divided into three segments - duodenum, jejunum and ileum (McLELLAND, 1986). Its characteristic mucosa is full of villi (DUKE, 1996), where the main phase of chemical digestion (POUGH et. al., 1999) and nutrient absorption take place (HARCOURT-BROWN & CHITTY, 2005).

In psittacines, the small intestine is relatively simple. The first portion of the intestine is the duodenum (SCHMIDT et. al., 2003). It is a loop made up of a proximal descending portion and a distal ascending portion (McLELLAND, 1986), forming a closed curve in a "U" shape (NICKEL et. al., 1977; DYCE et. al., 1997) and involving the pancreas. Most of the loop is on the abdominal floor and follows the caudal curvature of the gizzard (DYCE et. al., 1997). Two shorter portions follow. The middle third of the second portion contains the rest of the yolk sac: the yolk diverticulum. It is considered the junction between the jejunum and ileum (SCHMIDT et. al., 2003), since there is a marked histological distinction between the jejunum and ileum (DUKE, 1996), although this is of little physiological importance. The ileum is then made up of two loops of similar length (SCHMIDT et. al., 2003)The large intestine is the great differentiator in psittaciformes, due to the aforementioned absence of caecums (or functional vestigial caecum). It is basically made up of a short, straight, continuous intestine with an ileum and cloaca (McLELLAND, 1986). The rectum is a short segment whose function is to accumulate faeces (SCHWARZE, 1980).

CHAPTER 6

Reproductive physiology

The natural history of reproduction in birds is unique and deserves special attention. They are obligately oviparous animals (they lay eggs), and are the only class of vertebrates in which there is no species that reproduces by viviparity (where the embryo develops inside the mother's body, in a placenta that provides it with the nutrients necessary for its development) (BLACKBURN & EVANS, 1986), i.e. birds ovulate a single oocyte at frequent intervals and must produce a fertile egg with all the needs of the developing embryo, without any subsequent direct maternal interference (FROMAN et.al, 2004).

6.1. Females

Birds have physiological adaptations, and in many species the female has only one ovary (usually the right one is atrophied, with the exception of Apterygiformes (Kiwi - *Apteryx sp.)).)* (FEDUCCIA, 1996), thus reducing body mass and facilitating flight, as in cockatiels, which only have a well-developed left ovary and oviduct (JOHNSON, 2000; POLLACK & OROSZ, 2002), which has five distinct areas: infundibulum, magnum, isthmus, uterus (shell gland), vagina (Figure 11)(KING & MCLELLAND, 1984; JOHNSON, 2000). The blood supply to the ovary is large and the blood vessels are short. The ovary is closely attached to the body wall, next to the adrenal gland and the cranial division of the kidney. It has many follicles, which are small in immature birds and enlarged by the presence of yolk in adult birds when they enter reproduction (HARCOURT-BROWN & CHITTY, 2005).

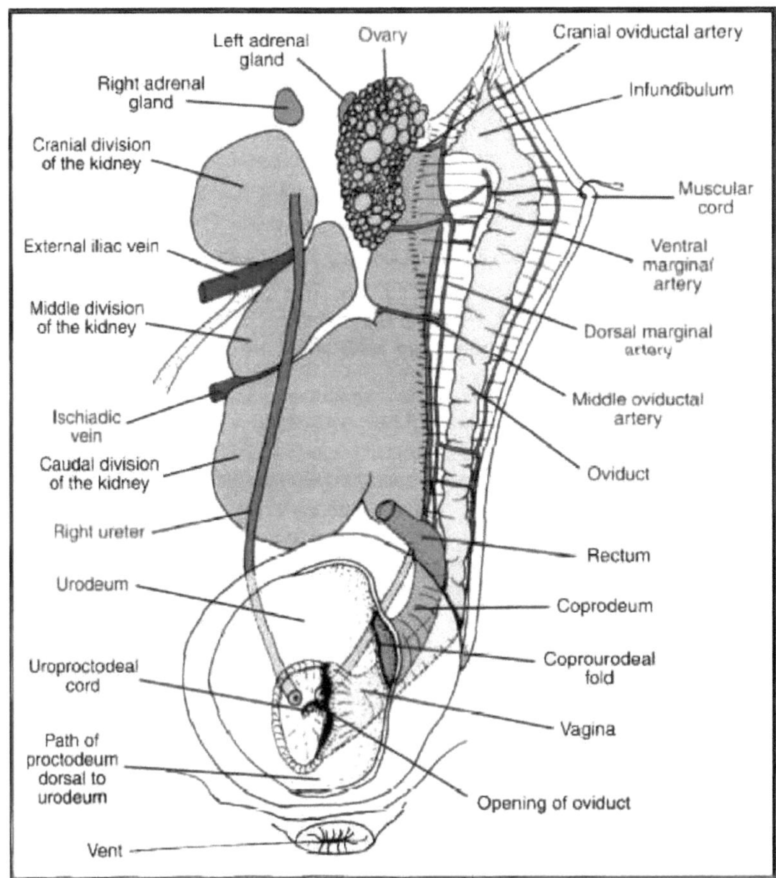

Figure 11. A view of the urogenital tract of a sexually mature female grey parrot *(Psittacus erithacus)*. Source: HARCOURT-BROWN & CHITTY, 2005.

Birds only have one follicular phase and do not need to form a corpus luteum like mammals, but the follicular growth cycle and the endocrine fluctuations that lead to ovulation are called the ovulatory cycle (PROUDMAN, 2004). It can be said that they have an ovarian follicular hierarchy, with the largest follicle ovulating at regular time intervals.

GnRH is a hormone produced by the hypothalamus with the function of regulating reproduction in vertebrates. Nerve stimulation results in the activation of the hypothalamic-pituitary gate system with the release of GnRH pulses that stimulate the secretion of follicle stimulating hormone (FSH) and luteinising hormone (LH) from the anterior pituitary gland (HAFEZ et al., 2004). It is the release of GnRH by the hypothalamus that initiates the laying process in birds. About 6 hours before a bird ovulates, she has a GnRH surge, which results in an LH spike. If there is a mature pre-ovulatory follicle, it will respond to this LH surge and produce progesterone. The progesterone will stimulate the release of more LH and so on. This interaction between LH and progesterone is fuelled by positive feedback (POLLOCK & OROSZ, 2002). Ovarian follicles that do not ovulate or do not complete their development undergo a process of follicular atresia, which can occur in all stages of follicular growth (BACHA & BACHA, 2003).

According to HARCOURT-BROWN & CHITTY (2005), the follicle develops suspended in a shaft of smooth muscle, blood vessels and nerves. It contains a large primary oocyte surrounded by a multi-layered wall, potentially divided by the stigma. At ovulation the stigma is divided; the secondary oocyte is 'grabbed' by the infundibulum. If the oocyte is shed, but not enveloped by the infundibulum, it is usually reabsorbed.

After the oocyte is shed from the ovary, the infundibulum surrounds it with a first layer of albumin (egg white). Fertilisation of the oocyte by the sperm is limited to a period of around 15 minutes, being the period between the release of the oocyte from the ovary and its filling with albumin. This is only possible because the spermatozoa are contained in folds in the mucosa of the oviduct.

The egg passes through the magna, where the rest of the endosperm is added. In the isthmus, the shell membranes are produced, covering the egg and determining its final shape. The egg then enters the shell gland, where it spends 80 per cent of its time in the oviduct. Firstly, the egg absorbs water and enlarges, acquiring a rounded shape. Next, the calcium carbonate and protein shell is produced and is finally covered with a cuticle giving the egg its shiny appearance. The shell contains thousands of tiny pores that run from the shell membranes to the surface, allowing the embryo to breathe.

The final part of the oviduct is the muscular vagina, which in many species has a spermatic fossula capable of storing sperm. In many species (e.g. *Budgerigar*) viable sperm can be stored and released for several days.

The transit time in the oviduct varies between species, but is typically around 24 hours (RITCHIE et. al., 1995). Thus, the female lays one egg per day during the reproductive season (UBUKA & BENTLEY, 2011), or lays her eggs on alternate days (HARCOURT-BROWN & CHITTY, 2005).

The incubation period of the eggs (brooding) is controlled by the hormone prolactin. In females that fail to incubate their eggs, the prolactin values found never exceed the baseline values. Similarly, in altricial species (immature chicks that are extremely dependent on parental care), prolactin values are high during incubation and after hatching, during the chick feeding period (GOLDSMITH, 1982). The prolactin peak coincides with the decline in LH concentration and gonadal regression in many bird species (SHARP & SREEKUMAR, 2001).

In a study by Myers et al, (1989), plasma levels of luteinising hormone (LH) and prolactin (PRL) were determined by radioimmunoassay during two reproductive cycles in captive cockatiels *(Nymphicus hollandicus)*. LH levels were elevated during nest inspection time in females and peaked during egg laying.

LH levels were also higher in males during nest inspection, but were lower during egg laying. In both sexes, LH continued to decrease during incubation and chick care. Prolactin levels in both sexes increased during egg laying and peaked during incubation. In other words, nest inspection and egg laying in cockatiels are characterised by high LH levels, while high PRL levels occur during incubation and chick feeding in both sexes.

6.2. Males

The male reproductive organs are a pair of equally active testes, next to each adrenal gland in the cranial division of the kidneys. In many parrots, the right testicle is smaller than the left, because the germ cells migrate from the right gonad (ovary or testes) to the left during development. Inactive testicles are small with no blood supply. When sexually active, they are much larger and have well-developed blood vessels. Sperm is collected in the epididymis and passes into the duet deferens, which expands distally into a sperm receptacle, the seminal accumulation. In this structure the temperature can be 6°C lower than that of the surrounding tissues. There are no accessory genital glands (HARCOURT-BROWN & CHITTY, 2005).

In most species, males do not have an intromittent organ (penis) of any kind, which could also be considered an adaptation for weight reduction during flight. Because of this lack of external genitalia, copulation in many birds involves brief cloacal contact, in which the spermatozoa are transferred from the male to the female (BALL & BALTHAZART, 2002).The action of gonadotropins in males is very similar to the mechanisms observed in mammals. Luteinising hormone stimulates Leydig cells to produce testosterone and androstenedione, while follicle-stimulating hormone acts on Sertoli cells, the mechanism of action of which is still unknown, but it is known that its action is potentiated by testosterone (CROSTA et al., 2003). This production of testosterone stimulates spermatogenesis, the growth of the epididymis and the development of the tubules, especially the vas deferens (POLLOCK & OROSZ, 2002). Another function of testosterone is attributed to the control of sexual behaviour involving the establishment and defence of territories or nests during the reproductive season, as well as singing and aggression (POLLOCK & OROSZ, 2002).

6.3. Advances in wild bird reproduction

One of the most important aspects in improving the reproductive efficiency of species is knowledge of the endocrine processes related to reproduction. This isolated knowledge of reproductive *status* can be a good indicator of individual welfare, but when integrated with other knowledge it can become an important tool in captive management (PICKARD, 2003). Endocrine steroid monitoring can be used to assess changes in the reproductive cycle, such as the onset of puberty, duration of sexual receptivity, seasonality, onset of reproductive senescence and the impact of environmental factors on the reproductive health of species (DONOGHUE et al., 2003).

Most of the accumulated knowledge about the endocrine physiology of poultry reproduction has been obtained from studying domestic birds, mainly chickens, turkeys, quails and geese. This is because traditional endocrine monitoring studies involve repeated blood sampling and are therefore more accessible in domestic animals, due to the stress generated, risk of accidents and availability of animals (CHRISTOFOLETTI, 2014).

In recent decades, there has been substantial progress in the study and understanding of the reproductive endocrinology of wild birds (ELPHICK et al., 2007), a fact related to the growing use of non-invasive endocrine monitoring techniques. Currently, the main uses of these techniques are the measurement of urofaecal metabolites of steroid hormones to characterise endocrine profiles, access reproductive *status* and identify reproductive disorders in wild birds (WASSER & HUNT, 2005; JENSEN & DURRANT, 2006; STALEY et al., 2007; PEREIRA, 2008).

This progress in the endocrinology of wild birds has been accompanied by hormonal manipulation to inhibit or stimulate reproduction (ELPHICK et al., 2007). To stimulate reproduction in captivity, hormone therapy with GnRH has achieved positive results (JAWOR et al., 2006; ROBBE et al., 2008; CONSTANTINI et al., 2009; ELNAGAR, 2009).

The administration of bird-specific GnRH or mammalian GnRH analogues has stimulated reproductive activity in different species of birds, producing an increase in plasma LH, gonadal development, changes in sexual behaviour and the production of young in the non-reproductive season (STERLING & SHARP, 1984; MINOIA et al., 1984; CONSTANTINI et al., 1985; McNAUGHTON et al., 1995; CONSTANTINI et al., 2009). In wild birds, the results of GnRH stimulation can be influenced by the reproductive season, especially in terms of photoperiod and photorefractoriness. The subcutaneous application of lecirelin (an analogue of mammalian GnRH) to Belgian canaries (*Serinus canaria*) outside of their reproductive season promoted the first laying much earlier, but a second laying only occurred with couples that were subjected to an artificial photoperiod corresponding to the reproductive season, demonstrating the dependence of reproductive activity on light stimulation (CONSTANTINI et al, 1985). The same happened in Juncos *(Junco hyemalis), where* testosterone levels were significantly elevated when subjected to intramuscular application of c-GnRH during the breeding season, with the response to the drug decreasing as the non-breeding season approached (JAWOR et al., 2006).

There are currently few reports of studies with psittaciformes relating to the effects of using GnRH on the reproduction of this family. In a study with the Australian parakeet *(Melopsittacus undulatus),* the effect of using a slow-release buserelin acetate implant on the reproductive activity of this species was evaluated through endocrine monitoring and follow-up of reproductive parameters (laying rate and fertility) (CONSTANTINI et al., 2009). The couples that received the implant had higher laying rates, a higher proportion of egg fertility and a higher concentration of excreted steroid levels, but without modifying the typical excretion profiles when compared to the other couples that did not receive the implant.

The environment is a very important factor in controlling the reproductive cycle of birds. It can influence the time at which the bird will reproduce, including variations such as photoperiod, environmental temperature, access to nesting sites, food availability and a variety of social interactions between specimens (BALL & BALTHAZART, 2002), but the main one involved in controlling bird reproduction is photoperiod (SHARP et. al, 1998; JOHNSON, 2006), especially in temperate regions where the seasons are well defined.

The photoperiod is responsible for synchronising and determining a reproductive dynamic for birds throughout the year, establishing an annual sexual cycle (GOODSON et. al, 2005; BARALDI-ARTONI et. al, 2007). Through photoperiodism, birds will restrict their reproduction to times when environmental conditions are favourable.

Many authors believe that the hypothalamus-pituitary-gonads axis responds to the increased amount of daylight after the winter solstice. This generates an increase in the secretion of gonadotropins, triggering an increase in the gonads and a wide range of steroid hormone-dependent processes, including changes in reproductive behaviour. Birds that become reproductively active during the stimulus of an increase in the amount of daily light are called photostimulable, while those that do not respond to an increase in the amount of daily light are called photorefractory (BALL & BALTHAZART, 2002).

There are species whose reproduction is stimulated by a decrease in photoperiod, such as the caulker *(Padda oryzivora)* (SAITO *et al.*, 1992) and the emu *(Dromaius novaehollandiae)* (BLACHE *et al.*, 2001). However, bird species that breed naturally in the tropics, where they do not experience significant variations in the amount of daily light, reproductive variations are less influenced by photoperiod (VLECK, 1993).

In birds, light is perceived thanks to photoreceptors that transform the energy contained in photons into biological signals. The energy of the photons in the eye is transformed by the photosensitive pigments contained in the cones and rods and transmitted by neurons to the brain, where the signal is integrated into an image (JÁCOME, 2009). Light is perceived by hypothalamic photoreceptors, which convert the electromagnetic signal into a hormonal message through its effects on hypothalamic neurons that secrete gonadotrophin-releasing hormone (GnRH). GnRH acts on the pituitary gland to produce the gonadotropins: luteinising hormone (LH) and follicle stimulating hormone (FSH). LH and FSH bind to their receptors on the theca and granulosa cells of the ovarian follicle, stimulating the production of androgens and estrogens by the small follicles and the production of progesterone by the larger pre-ovulatory follicles. Short days don't stimulate adequate gonadotrophin secretion because they don't illuminate the entire photosensitive phase. Longer days, however, do stimulate, and thus LH production is initiated. This neurohormonal mechanism controls the reproductive and behavioural functions and secondary sexual characteristics of birds (ROCHA, 2008).

According to Rocha (2008), for breeding purposes, however, the perception of light in birds does not depend solely on the photoreceptors in the eye. Light needs to pass through the bones of the skull to stimulate specific photoreceptors in the hypothalamus. Contrary to what is observed in mammals, the perception of light information in birds is more important via the "transcranial" route than the ocular route, as it was shown in their study that darkening the skull of sparrows *(Passer domesticus)* with India ink blocks the sexual response on long days, while deprivation of the eyeball did not have the same effect. It's possible that the eyes are not indispensable to light stimulation but play a synchronising role in circadian rhythms - sleeping and waking.

Variations in environmental temperature can also modulate responses in the hypothalamus-pituitary-gonads axis, although they are quite subtle factors when studied experimentally. In many cases, temperature variations are clearly seen as a factor influencing the timing of egg laying under natural conditions in females of various species (MEIJER *et al,* 1999).

The availability of food is a significant factor for reproduction (SCHEUERLEIN & GWINNER, 2002). However, in many species, it cannot be considered a stimulating factor for the development of the reproductive system. This does not mean that feeding is not fundamental, but rather that it is more closely related to the nutritional condition of the female, which can influence egg laying (MEIJER *et al,* 1999). In the tropics, most of the bird species studied tend to have a breeding season that is usually related to the annual rainfall pattern, when there is plenty of food (HAU et. al, 2004).

Social interactions are also very important for stimulating reproduction. It is believed that one of the most important aspects of reproduction is for the female to have a safe place to lay her eggs (nest) in the presence of the male. The courtship ritual of males towards females has a strong effect on the endocrine physiology of females. Studies with the Australian parakeet *(Melopsittacus undulatus) have* shown that male singing can stimulate ovarian growth in females (BROCKWAY, 1969). Similarly, the female also plays a fundamental role in the reproductive physiology of males. A study of birds of the common starling species *(Sturnus vulgaris)* showed that the growth of the gonads in males is greatly increased if they are housed together with females (BURGER, 1953). In semi-mated pigeons (part of the testicles removed), the presence of the female in the enclosure can have an influence and can cause compensatory hypertrophy in these testicles (CHENG, 1974; 1976).In some psittaciformes such as cockatiels *(Nymphicus hollandicus)* (MYERS *et al.,* 1989; SHIELDS *et al.,* 1989), photostimulation works in conjunction with easy access to the nest and the presence of the male to promote greater release of LH secretion and reproductive activity, so that this species breeds in captivity all year round.

CHAPTER 7

Reproductive management

A better understanding of reproductive behaviour can increase the success of captive breeding programmes for endangered psittacine species. One third of all psittacine species are considered endangered (COLLAR & JUNIPER, 1992).Calopsitas *(Nymphicushollandicus)* are birds with visible sexual dimorphism, in which the males have a bright yellow head and vibrant orange cheeks, and the females have a less intense head colour and restricted yellow colour in the tail, with the presence of horizontal stripes on the tail (Figure 12). However, these characteristics are only observed in the original colour, wild grey (or wild grey), and are not valid for other existing colour mutations. Males sing and show courtship activities, characteristics that are not observed in females (ZANN, 1965; SMITH, 1978).

Figura 12. Male wild grey cockatiel (left) and female wild grey cockatiel (right). Source: http://www.produtosaguia.com.br/img/calopsita.jpg

They are monogamous birds, but it is also possible that non-exclusive pairing can occur in this species when bred in captivity. Seibert & Crowell-Davis (2001) observed in their work that in a captive group of cockatiels, both males in a polygamous trio looked after the female, although she was only observed copulating with one of them, and a male who copulated with two females subsequently looked after the chicks of both females, but in two different nest boxes.

When in their natural *habitat*, they nest in trees or cavities near water, with both sexes taking part in incubation and parental care (FORSHAW & COOPER, 1981). Breeding in captivity is stimulated by providing nest boxes and photostimulation (MILLAM et. al, 1998). Psittacines accept various forms of nest. They can be of the box type, brass, barrel, etc. The substrate that will line the bed for the eggs is prepared with tree trunk chips, wood shavings or autoclaved sand. You should choose the nest and substrate that best suits the conditions of the breeding site (ABRAMSON *etal.,* 1995).

According to Allgayer & Cziuli (2007), birds will only breed if the management meets their basic ethological needs. High nests and perches help convey security to the pair. Slippery perches make copulation difficult and can result in a large number of infertile eggs. Interference from strange noises, people or even dominant animals of the same species in visual contact, inhibit the pair from copulating or laying.

The pair, sure of their territory, exchange food, the male increases his aggression and they begin to copulate. Due to the absence of external genitalia in the male, copulation occurs through brief cloacal contact, in which the spermatozoa are transferred from the male to the female (BALL & BALTHAZART, 2002).

The eggs are laid in approximately 4 to 8 white-coloured eggs because, as they hatch in cavities, there is no need to camouflage the eggs, which makes it easier for the couple to handle them inside the dark nest. Incubation usually begins when the second egg is laid, ranging from 19 to 28 days (ALLGAYER & CZIULIK, 2007). After the 10th^Q day of incubation, it is possible to perform an ovoscopy, which consists of placing the egg against the light to confirm the presence or absence of an embryo (HARVEY, 1990).

There is a strategy used in captivity, called replacing lost eggs, which can be used by breeders to increase egg production by 100 per cent in some cases. The practice consists of removing the first egg when the bird lays the second, and so on. In this way, it is possible to reproduce five chicks or more (ALLGAYER & CZIULIK, 2007). Removing eggs from the nest in captivity stimulates new laying and increases egg production (FRANCISCO, 2012).

After the incubation period, when the chick hatches from the egg, both the female and the male feed it through regurgitation, forming a gruel that facilitates the return of the food (LOCATELLI et.al, 2013). As soon as the chicks are able to feed independently, they can be transferred to large cages (nurseries) where they can exercise their flight, socialise with other species or be sold (SWEENEY, 2000; ROMAGNANO, 2005; TSCHUSDIN et al., 2010).

If the parents do not provide parental care, the chicks can be hand-reared, which has several advantages, including the production of more docile birds, an increase in a couple's production during the breeding season, the opportunity to rescue sick, abandoned or high genetic value offspring, and a reduction in the burden of parental care when the couple has a large clutch. The disadvantages of hand rearing must also be considered, such as the intensive work that requires specialised labour and technology, the easier spread of pathological outbreaks, the influence on their natural behaviour as adult birds and the fact that hand reared birds rarely gain weight more quickly than those reared by their parents (TSCHUDIN et al. 2010).

The puppy is the primary product of a commercial breeding or conservation programme for a species. You should always keep an individual record of each chick, which includes: date of birth, sex, identification of parents, ring and/or microchip number. This record should include the monitoring and evaluation of growth (ALTMAN et al., 1997; RUPLEY, 1999). This cub monitoring data is essential for establishing the parameters of the species' development and can serve as a basis for breeding other species in the future.

12.1.Artificial incubation

Artificial incubation is commonly used by commercial breeders. Its advantages include an increase in the number of eggs per couple and consequently a greater number of chicks and the possibility of continuous control over the development of the embryo and later the chicks (ALLGAYER & CZIULIK, 2007). Little is known about the artificial egg incubation technique for this species, and in practically all breeding centres the chicks are incubated by the parents and remain with them until they become independent or are sold. Currently, one of the biggest problems faced by cockatiel breeders is the lack of information, which jeopardises the management and reproduction of the *Nymphicus hollandicus* species.

7.2 Artificial insemination

Although the diversity of the world's birdlife is quite significant, the extinction of many species is a reality. According to Myers (1987) the wave of extinction of living species caused by human pressure and exploitation is occurring at a rate 400 times greater than the natural rate. Consequently, this pressure is hitting birds hard. And one of the important factors in this pressure, apart from the destruction of *habitats,* is trafficking.

Captive breeding is important both for maintaining biological diversity, which should be used in conservation programmes, and for commercial purposes, to meet the demand for pet birds. These objectives are interconnected and can be complementary (ALLGAYER & CZIULI, 2007).

As the cockatiel has been considered an ideal model species for studying the behaviour of psittaciformes (YAMAMOTO et al., 1989), Neumann et al, (2013), in their study, evaluated artificial insemination in 32 cockatiels (*Nymphicus hollandicus),* divided into two groups. In one group, the males were sterilised endoscopically. The males in the other group were used as semen donors. After collection using a new massage technique (Figure 13), semen samples were examined under a microscope to assess contamination and quality.

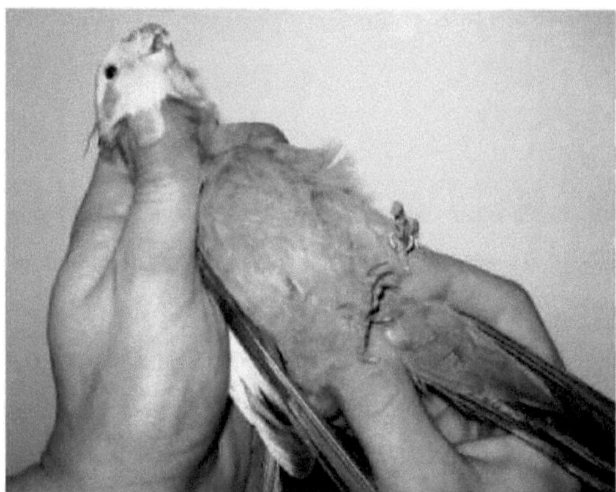

Figura 13.　　Massage technique for semen collection in a male cockatiel *(Nymphicus hollandicus)*. Source: NEUMANN et al., (2013).

Samples with medium to high sperm concentrations (Figure 14), medium to high motility and no contaminants were used for intracloacal artificial insemination of cockatiels (Figure 15) in the group with sterile males. In total, 74.2 per cent of all attempts to collect semen were successful. Insemination resulted in the fertilisation of 17 out of 23 eggs (73.9%), which was slightly lower than the natural fertilisation rate (88.4%). Easily applicable in veterinary practice, this study demonstrates that the use of artificial insemination can be a valuable tool for dealing with reproductive failures in captive psittaciformes.

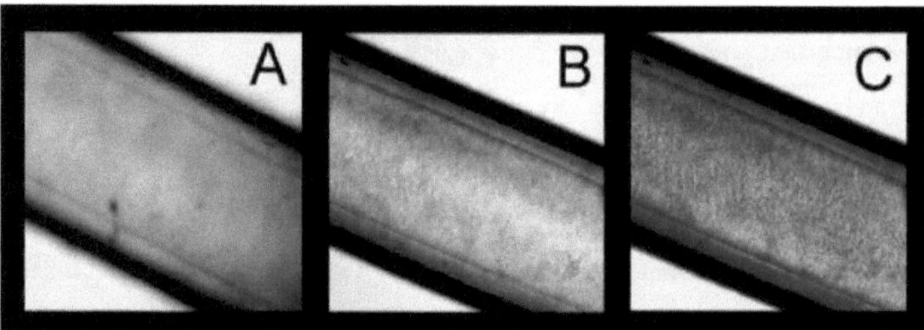

Figura 14.　　Estimated sperm concentrations (A = low, B = C = medium, high) in semen from a male cockatiel *(Nymphicus hollandicus)*. Source: NEUMANN et al., (2013).

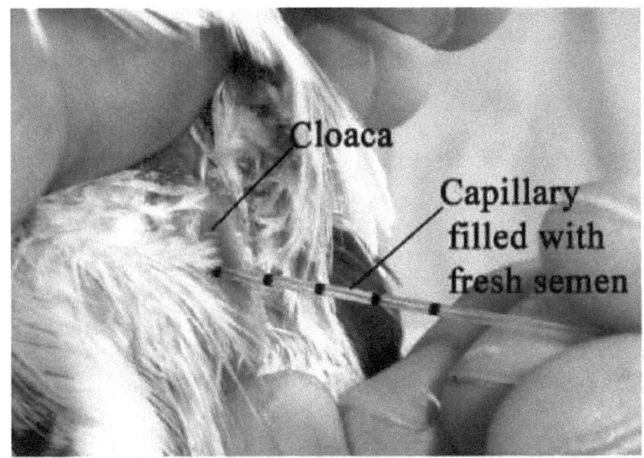

Figura 15.   Intracloacal insemination of a female cockatiel *(Nymphicus hollandicus).*

Source: NEUMANN et al., (2013).

1.3. Egg retention

Egg retention is the inability of an egg to pass through the oviduct at a normal speed (delayed oviposition). The main symptoms presented by birds with egg retention are: apathy, accelerated respiratory movements, distended lower oviduct and reluctance to fly. The bird lies weak and powerless on the floor of the cage. Egg retention can be detected by palpating the abdomen, which becomes prominent, or through X-ray examinations (Figure 16) (DANTE, 2008).

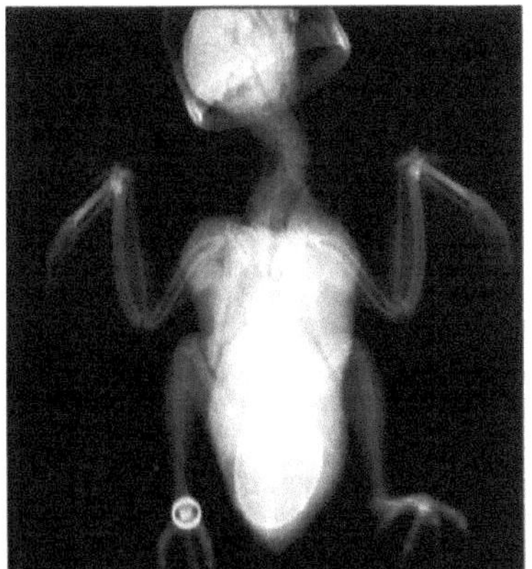

Figure 16. X-ray of a retained egg in a psittacine, showing a normal-sized egg surrounded by a shell. The bird was given calcium supplementation and the egg was laid within half an hour. The egg had a thinner shell than normal and was missing a cuticle (probably hypocalcaemia). Source: HARCOURT-BROWN & CHITTY, 2005.

The main causes of egg retention are: genetic tendency; poor nutrition (low calcium concentration); stress; large eggs; eggs with shell deformities (thin, missing, altered shape); older females and excessive laying. Thus, whatever the reason, it is always a delicate situation and can happen at any time during laying, not necessarily with the first egg (DANTE, 2008).If egg retention is diagnosed, the ideal course of action is to administer oxytocin to the female in an attempt to stimulate laying, thus increasing the number of contractions of the oviduct and facilitating the passage of the egg. If laying doesn't occur, the retained area should be massaged in an attempt to remove the egg without invasive methods. If laying still does not occur, surgery to remove the egg should be carried out, as egg retention can lead to the animal's death (LEITE, 2011).

CHAPTER 8

Final considerations

Behavioural studies of the cockatiel species *Nymphicus hollandicus* are scarce in the literature. A better understanding of their behaviour in captivity is crucial for a better understanding of their physical and social needs, thus contributing to proper management in commercial breeding facilities, since the cockatiel is one of the most widely bred species of *Psittaciformes in* the world as a pet bird, as well as other species of *psittaciformes,* including endangered birds, thus aiding in fauna conservation projects and the breeding of psittaciformes in zoos. In addition, captive animals depend exclusively on humans for their survival, so animal welfare becomes an indispensable element for a satisfactory quality of life in captivity.

References

ABRAMSON, JSPER, B. L.; THOMSEN, J. B. T. The large macaws. Fort Bragg, CA: Raintree Publications, 1995.

ABREU, P.G.; ABREU, V.M.N.; MAZZUCO, H. Use of evaporative (adiabatic) cooling in broiler rearing. Concórdia: Embrapa Swine and Poultry, 50 p. 1999. Available at :<http://www.infoteca.cnptia. embrapa. br/bitstream/doc/437183/l/doc59.pdf>.
Accessed on: 19 June 2015.

ABREU, P. G.; ABREU, V. M. N. Lantern: function and construction. Embrapa Suínos e Aves, 2p. 2000a. Available at: <http://ainfo.cnptia.embrapa. br/digital/bitstream/CN PSA/15608/l/itav015.pdf>. Accessed on: 19 June 2015.

ABREU, P.G.; ABREU, V.M.N. Ventilação na avicultura de corte. Concórdia: Embrapa Swine and Poultry, 50p. 2000b. Available at :< http://ainfo.cnptia.embrapa.br/digital/bitstream/item/58306/l/doc63.pdf>. Accessed on: 19 June 2015.

ABREU, V. M. N.; ABREU, P. G. D. The challenges of ambience on poultry systems in Brazil. Revista Brasileira de Zootecnia, v. 40, p. 1-14, 2011.

ALLGAYER, M.; CZIULIK, M. Reproduction of psittaciformes in captivity. Revista Brasileira de Reprodução Animal, v. 31, p. 344-350, 2007.

ALTMAN, R.; CLUB, ,L.; DORRESTEIN, G. M.; QUESENBERRY, K. Psittacine paediatric husbandry and medicine. Avian Medicine and Surgery. I[rd] ed. WB Saunders, p. 73-95, 1997.

ANGELO, M. S. P.; NAAS, I.; VENDRAMETTO, O. Computer programme for estimating thermal comfort in intensive pig and broiler production. Engineering in Agriculture, v. 22, n. 6, p. 535-542, 2014.

ANTONIALLI, L.M.; SOUKI, G.Q.; TEIXEIRA, T. H. Strategies for the commercial breeding of wild birds: The case of a rural enterprise authorised by IBAMA. In: XLII CONGRESSO BRASILEIRO DE ECONOMIA E SOCIOLOGIA RURAL, 16p. 2004. Available at: <http://www.sober.org.br/palestra/12/080382.pdf>. Accessed on: 12 June 2015.

ARCARO JÚNIOR, L; ARCARO, J. R.; POZZI, C. R.; FAGUNDES, H.; MATARAZZO, S. V.; OLIVEIRA, C. A. D. Plasma levels of hormones, production and composition of milk in an air-conditioned waiting room. Revista Brasileira de Engenharia Agrícola e Ambiental, v.7, n.2, p.350-354, 2003.

AUSTRALIAN MUSEUM. Birds in backyards. Cockatiel *(Nymphicus hollandicus).* 2006. (Provides information on Australia's native birds). Available at: <http://www.Birdsinbackyards.Net/bird/49>. Accessed on: 22 June 2015.

AVILA, V.S.; KUNZ, A.; BELLAVER, C.; PAIVA, D.P.; JAENISCH, F.R.F.; MAZZUCO, H.; TREVISOL, I.M.; PALHARES, J.C.P.; ABREU, P.G.; ROSA, P.S. Good broiler production practices. Concórdia: Embrapa Swine and Poultry, 28p. 2007.

BACHA, W.J.; BACHA, L.M. Colour Atlas of Veterinary Histology. São Paulo, Brazil: Roca. 2rd ed., Roca, São Paulo, Brazil, 457p. 2003.

BAÊTA, F. C. Planning poultry facilities considering temperature variations. In: International Symposium on Ambience and Facilities in Industrial Poultry Production, Campinas. Proceedings. Campinas: FACTA, p.123-129.1995.

BAETA, F. C.; SOUZA, C. F. Ambiência em edificações rurais: conforto animal. Viçosa: UFV,. 246p. 1997.

BALL, G. F.; BALTHAZART, J. Neuroendocrine mechanisms regulating reproductive cycles and reproductive behaviour in birds. Hormones, brain and behaviour, v. 2, p. 649-798,2002. BARALDI-ARTONI, S. M.; BOTTINO, F.; OLIVEIRA, D.; FRANZO, V. $.; AMOROSO, L; ORSI, A. M.; CRUZ, C. Morphometric study of Rynchotus rufescens testis throughout the year. Brazilian Journal of Biology, v. 67, n. 2, p. 363-367, 2007.

BENNETT A, R.; DEEM, S. L. The gastrointestinal system of birds: I. Compendium of Continuing Education for the Veterinarian, Florida, v. 1, n. 1, p. 50-56,1996.

BELLAY, T.; TEETER, R. G. Broiler water balance and thermobalance during thermoneutral and high ambient temperature exposure. Journal Poultry Science, v. 72, n. 2, p. 116-124, 1993.

BENEZ, S. M. Aves: criação, clínica, teoria, prática. São Paulo: Robe Editorial, 522 p. 2001. BERNSTEIN, I. S. Dominance relationships and ranks: Explanations, correlations, and empirical challenges. Behavioural and brain Sciences, v. 4, n. 03, p. 449-457,1981.

BIRDLIFE INTERNATIONAL 2012. *Nymphicus hollandicus.* The IUCN Red List of Threatened Species. Version 2014.3. Available at: <http://www.iucnredlist.org>. Accessed on: 22 June 2015.

BLACHE, D.; TALBOT, R. T.; BLACKBERRY, M. A.; WILLIAMS, K. M.; MARTIN, G. B.; SHARP, P. J. Photoperiodic control of the concentration of luteinizing hormone, prolactin and testosterone in the male emu *(Dromaius novaehollandiae),* a bird that breeds on short days. Journal of Neuroendocrinology, v. 13, p. 998-1006, 2001.

BLACKBURN, D. G.; EVANS, H. E. Why there are no viviparous birds. The American Naturalist, v.128, p. 165-190,1986.

BONNET, S.; GERAERT, P. A.; LESSIRE, M.; CARRE, B.; GUILLAUMIN, S. Effect of high ambient temperature on feed digestibility in broilers. Journal Poultry Science, v. 76, n. 6, p. 857-863, 1997.

BORGES, S. A.; MAIORKA, A.; SILVA, A. V. F. Physiology of heat stress and electrolyte utilisation in broiler chickens.Ciência Rural, v. 33, n. 5, p. 975-981, 2003.

BRAZIL Ministry of the Environment, Water Resources and the Legal Amazon. Brazilian Institute of the Environment and Renewable Natural Resources. Ordinance n^9 93 of 07 July 1998 - Legislation on Brazilian fauna. Available at: <www.ibama.gov.br/fauna/legislacao/port_93_98.pdf>. Accessed on: 22 June 2015.

BRIDI, A. M. Animal adaptation and acclimatisation. Londrina: Department of Zootechnics. 15p. 2006. Available at   : <http://www.uel.br/pessoal/ambridi/Bioclimatologia_arquivos/AdaptacaoeAclimatacao   Animal.pdf>. Accessed on: 02 June 2015.

BROCKWAY, B. F. Roles of budgerigar vocalisation in the integration of breeding behaviour. Bird vocalisations, p. 131-158,1969.

BROOM, D. M. The Biology of Behaviour. School Science Review, v. 62, n. 220, p. 442-51, 1981.

BROOM, D. M.; JOHNSON, K. G. Stress and animal welfare. London: Chapman and Hall, 217p. 1993.

BUFFINGTON, D. E.; COLLAZO-AROCHO, A.; CANTON, G. H.; PITT, D. Black globe-humidity index (BGHI) as comfort equation for dairy cows. Transactions of the ASAE [American Society of Agricultural Engineers](USA), v. 24, n. 3, p. 711-714,1981.

BURGER, J. W. The effect of photic and psychic stimuli on the reproductive cycle of the male starling, Sturnus vulgaris. Journal of Experimental Zoology, v. 124, n. 2, p. 227-239, 1953.

CHAMP, M.; SZYLIT, O.; RAIBAUD, P.; ÁÍUT-ABDELKADER, N. Amylase production by three Lactobacillus strains isolated from chicken crop. Journal of applied bacteriology, v. 55, n. 3, p. 487-493,1983.

CHENG, M. F. Ovarian development in the female ring dove in response to stimulation by intact and castrated male ring doves. Journal of Endocrinology, v. 63, n. 1, p. 43-53,1974.

CHENG, M. F. Interection of lighting and other environmental variables on activity of hypothalamo-hypothpyseal-gonadal System. Nature, v. 263, p. 148-149,1976.

CHRISTOFOLETTI, M. D. Reproduction of true parrots *(Amazona aestiva)* in captivity: annual sex steroid profile and trial of exogenous hormonal stimulation. 72 p. 2014. PhD thesis - Universidade Estadual Paulista Júlio de Mesquita Filho, Faculdade de Ciências Agrárias e Veterinárias de Jaboticabal, 2014. Available at: <http://hdl.handle.net/11449/121900>. Accessed on: 02 June 2015.

COLLAR, N. J. Family *Psittacidae* (parrots). In: DEL HOYO, J.; ELLIOT, A.; SARGATAL, J. (Ed.). Handbook of the birds of the world: Sandgrouse to cuckoos. Barcelona: Lynx Edicions, v. 4, p. 280-447,1997.

COLLAR, N. JJUNIPER, A. T. Dimensions and causes of the parrot conservation crisis. New World parrots in crisis: Solutions from conservation biology, p. 1-24,1992.

COSTA, E. C. Arquitetura ecológica, condicionamento térmico natural. 5rd ed., São Paulo: Edgard Blúcher, 264p. 1982.

COSTANTINI, V.; CINONE, F.; LACALANDRA, G. Early oviposition in *Serinus canaria* after endonasal administration of Gn-RH.(Preliminary research)]. Bollettino delia Società italiana di biologia sperimentale, v. 61, n. 4, p. 633,1985.

COSTANTINI, V.; CARRARO, C.; BUCCI, F.; SIMONTACCHI, C.; LACALANDRA, G.; MINOIA, P. Influence of a new slow-release GnRH analogue implant on reproduction in the Budgerigar *(Melopsittacus undulatus,* Shaw 1805). Animal reproduction Science, v. 111, n. 2, p. 289-301, 2009.

CROSTA, L; GERLACH, H.; BÚRKLE, M.; TIMOSSI, L. Physiology, diagnosis, and diseases of the avian reproductive tract. Veterinary Clinics of North America: Exotic Animal Practice, v. 6, n. 1, p. 57-83, 2003.

CUBAS, Z. S; SILVA, J. C. R; DIAS, J. L. C. Tratado de Animais Selvagens - Medicina Veterinária. São Paulo: Roca, 1354p. 2006.

CURTIS, S.E. Environmental management in animal agriculture. AMES. The Iwoa State University, 409 p. 1983.

DAMASCENO, F. A.; SCHIASSI, L; SARAZ, J. A. O.; GOMES, R. C. C.; BAÊTA, F. D. C. Architectural concepts of facilities used for poultry production aiming at thermal comfort in tropical and subtropical climates. PUBVET, Londrina, v. 4, n. 42, Ed. 147, Art. 991, 2010.

DANTE, D. Problems with egg entrapment. Clube do criador. 2008. Available at: <http://www.clubedocriador.com>. Accessed on: 02 June 2015.

DESHAZER, J. A.; BECK, M. M. University of Nebraska Report for Northeast regional poultry project NE-127. Lincoln: Agricultural research Division, Univ. of Nebraska, 1988. DONOGHUE, A.; BLANCO, J. M.; GEE, G.; KIRBY, Y.; WILDT, D. Reproductive technologies and challenges in avian conservation and management. CONSERVATION BIOLOGYSERIES- CAMBRIDGE-, p. 321-337, 2003.

DUKE, G. E. Digestion in birds. Duke's Physiology of Domestic Animals. IIrd ed., Guanabara Koogan, São Paulo, 721p. 1996.

DYCE, K. M.; SACK, W. O.; WENSING, C. J. G. Tratado de anatomia veterinária. 2rd ed., Rio de Janeiro: Guanabara Koogan, p. 631-650,1997.

ELNAGAR, S. A. Response of Alexandria cockerels reproductive status to GnRH (Receptal) injection. International Journal of Poultry Science, v. 8, n. 3, p. 242-246, 2009.

ELPHICK, C. S.; REED, J. M.; DELEHANTY, D. J. Applications of reproductive biology to bird conservation and population management.Reproductive biology and phylogeny of aves (birds). Enfield (NH): Science Publishers. p. 367-399, 2007.

ESMAY, M. L. Principies of animal environment. 2rd ed., Westport: CT Abi, 325p. 1969. FEDUCCIA, A. The Origin and Evolution of Bird. Yale University Press, New Haven, CT and London, 1996.

FORD, J. Speciation in Australian Birds Adapted to Arid Habitats.Emu, v.74, p.161-168, 1974.

FORSHAW, J. M. Parrots of the World. 3rd ed., Landsdowne Press, Willoughby, New South Wales, Australia. 1989

FORSHAW, J.M. Parrots of the World. Princeton: Princeton University Press, 2010. 328p.

FORSHAW, J. M.; COOPER, W. T. Parrots of the World. 2rd ed., Melbourne: Lansdowne Press, 616p. 1981.

FRANCISCO, L. R.; MOREIRA, N. Management, reproduction and conservation of Brazilian psittaciformes. Revista Brasileira de Reprodução Animal, v.36, p.215-219, 2012.

FROMAN, D. P.; KIRBY, J. D.; PROUDMAN, J. A. Reproduction in birds: Male and female. Animal Reproduction. 7[3] ed. Manole, Barueri, p. 237-257, 2004.

FURLAN, R.L Influence of temperature on broiler production. Simpósio Brasil Sul de Avicultura, v. 7, n. 04, p. 104-135, 2006.

FURTADO, DA; TINÔCO, I.F.F.; NASCIMENTO, J.W.B.; LEAL, A.F.; AZEVEDO, M.A. Characterisation of poultry facilities in the agreste paraibano mesoregion. Engenharia Agrícola, v.25, n.3, p.831-840, 2005.

GATES, R. S.; ZHANG, H.; COLLIVER, D. G.; OVERHULTS D. G. Regional variation in temperature index for poultry housing. Transactions of the ASAE, St. Joseph, v. 38, n. 1, p. 197-205, 1995.

GREENACRE, C.B.; LUSBY, A.L Physiologic responses of Amazon parrots *(Amazona species)* to manual restraint. Journal of Avian Medicine and Surgery, v.18, n.l, p.19-22, 2004.

GRAHAM, J.; WRIGHT, T. F.; DOOLING, R. J.; KORBEL, R. Sensory capacities of parrots. Manual of parrot behaviour. 2rd ed., Luescher, A. U. Blackwell Publishing, 113 p. 2006.

GODOY, S.N. *Psittaciformes* (Macaw, Parrot, Parakeet). In: CUBAS, Z. S; SILVA, J. C. R; DIAS, J. L. C. Tratado de Animais Selvagens - Medicina Veterinária. São Paulo: Roca, 1354p. 2006.

GOLDSMITH, A. R. Plasma concentration of prolactin during incubation and parenteral feeding throughout repeated breeding cycles in canaries *(Serinus canarius)*. Journal Endocrinology, v. 94, p. 51-59, 1982.

GOODSON, J.L.; SALDANHA, C.J.; HAHN, T.P.; SOMA, K.K. Recent advances in behavioural neuroendocrinology: insights from studies on birds. Hormones and behaviour, v. 48, n. 4, p. 461-473, 2005.

GORMAN, M. The Cockatiel Handbook. Barron's Educational Series, 139 p. 2010.

GUYTON, A.C.; HALL, J.E. Treatise on Medical Physiology. Rio de Janeiro: Guanabara-Koogan, 160p. 2002.

HAFEZ, E.; JAINUDEEN, M.; ROSNINA, Y. Hormones, growth factors and reproduction. In: Hafez, E. and Hafez, B. (Ed.). Animal reproduction, v.7, p.33-54, 2004.

HARCOURT-BROWN, N.H. Psittacine birds. In: Tully, T.N.; LAWTON, M.P.C.; DORRESTEIN, G.M. Avian Medicine. Elsevier, p.114-5, 2003.

HARCOURT-BROWN, N.; CHITTY, J. BSAVA MANUAL OF PSITTACINE BIRDS. 2rd ed., British Small Animal Veterinary Association. 333 p. 2005.

HARDY, J.W. Flock social behaviour of the Orange-fronted Parakeet.Condor, p. 140-156, 1965.

HARDOIM, P.C. Facilities for dairy cattle. Brazilian Congress of Agricultural Engineering. National Meeting of Technicians, Researchers and Educators of Rural Constructions, Poços de Caídas: SBEA/UFLA, chap. 3, p. 149-208,1998.

HARVEY R. Practical incubation. Farenham: Paynessex Printers, 1990.

HAU, M.; WIKELSKI, M.; GWINNER, H.; GWINNER, E. Timing of reproduction in a Darwin's finch: temporal opportunism under spatial constraints. Oikos, v. 106, n.3, p. 489-500, 2004.

JÁCOME, I. M. T. Different artificial lighting systems used to house light layers. 120p. 2009. PhD Thesis - State University of Campinas, Faculty of Agricultural Engineering. Campinas, 2009.

JACKSON, W.M. Why do winners keep winning? Behavioural Ecology and Sociobiology, v. 28, n. 4, p. 271-276,1991.

JAWOR J. M.; MCGLOTHLIN J. W.; CASTO J. M.; GREIVES T. J.; SNAJDR E. A.; BENTLEY G. E.; KETTERSON E. D. Seasonal and individual variation in response to GnRH challenge in male dark-eyed juncos *(Junco hyemalis)*. General and comparative endocrinology, v. 149, n. 2, p. 182-189, 2006.

JENSEN T. AND DURRANT B. Assessment of reproductive status and ovulation in female brown kiwi *(Apteryx mantelli)* using faecal steroids and ovarian follicle size. Zoo Biology, v. 25, n. 1, p. 25-34, 2006.

JOHNSON, A. L. Reproduction in the Female. In: WHITTOW, G. C. Sturkie's Avian Physiology. San Diego: Academic Press, p. 569-596, 2000

JOHNSON, P.A. 2006. Poultry Reproduction, p. 691-701. In: Dukes, Physiology of Domestic Animals. 12rd ed., Guanabara Koogan, Rio de Janeiro.

JONES, D. Feeding ecology of cockatiel, *Nymphicus hollandicus,* in a grain-growing area. Australian Wildlife Research, Victoria, v. 14, n. 1, p. 105-115,1987.

KELLY, C.F. BOND, T.E. ITTNER, N.R. Cold spots in the skay help cool livestock. Agricultural Engineering, v.31, n.12, p.606-606,1950.

KING, A. S.; McLELLAND, J. Female reproductive system. In: Birds: their structure and function. 2rd ed., London: Baillière Tindal, p. 145-165,1984.

KOUTSOS, E. A.; MATSON, K. D.; KLASING, K.C. Nutrition of birds in the order Psittaciformes: a review. Journal of Avian Medicine and Surgery, v. 15, n. 4, p. 257-275, 2001a.

KOUTSOS, E.A.; SMITH, J.; WOODS, L.W.; KLASING, K.C. Adult cockatiels (Nymphicus hollandicus) metabolically adapted to high protein diets. The Journal of nutrition, v. 131, n. 7, p. 2014-2020, 2001b.

LEITE, A. I. Egg retention in a parrot (Amazona vinacea): Case report. 24p. 2011. Course Conclusion Paper - Formiga University Centre (UNIFOR-MG). 2011. LOCATELLI, A. C.; WRUBLACK, S. C.; BASILE, L. F.; DO

NASCIMENTO, A. F.; BERBER, G. D. C. M.; DE ARAÚJO BERBER, R. C. Reproductive and maternal behaviour of Canindé macaws *(Ara ararauna Linnaeus,* 1758) kept in captivity for conservation. Comunicata Scientiae, v. 4, n. 4, p. 316-323, 2013.

LORO PARQUE. A comment on parrot nutrition. 2009. Available at: <http://www.loroparque-fundacion.org/>. Accessed on: 02 June 2015.

MACARI, M.; FURLAN, R.L.; GONZALES, E. Avian physiology applied to broilers. Jaboticabal: FUNEP/ UNESP, 246 p. 1994.

MACARI, M., FURLAN, R.L. Ambiência na produção de aves em clima tropical. Piracicaba: FUNEP, v. 1, p. 31-87, 2001.

MACHADO, P. A. R.; SAAD, C. E. P. The future of feed for ornamental and wild birds in Brazil. Aves - Revista Sul Americana de Ornitofilia, Belo Horizonte, v. 3, p. 37-40, 2000.

MATHIAS, J. M. How to raise cockatiels. Globo Rural magazine, São Paulo, 2008. Available at: <http://revistagloborural.globo.eom/GloboRural/0,6993, EEC1257601-4530,00.html>. Accessed on: 10 June 2015.

McLELLAND, J. Digestive system of birds. In: GETTY, R. Sisson/Grossman: Anatomy of Domestic Animals. 5rd ed., Rio de Janeiro: Guanabara Koogan, p. 1445-1464,1986.

MCNAUGHTON, F.; DAWSON, A.; GOLDSMITH, A. A comparison of the responses to gonadotrophin-releasing hormone of adult and juvenile, and photosensitive and photorefractory European starlings, *Sturnus vulgaris.* General and comparative endocrinology, v. 97, n. 1, p. 135-144,1995.

MEIJER, T.; NIENABER, U.; LANGER, U.; TRILLMICH, F. Temperature and timing of egg- laying of European starlings. Condor, 101,124-132,1999.

MILLAM, J.; ROUDYBUSH, E.; GRAU, C. R. Influence of environmental manipulation and nestbox availability on reproductive success of captive cockatiels *(Nymphicus hollandicus).* Zoo Biology, v.7, p.25-34,1998.

MINOIA, P.; DE BENEDICTIS, G.; LACALANDRA, G.; LATERZA, V.; BUFO, P. Reproductive conditioning of the partridge *(Perdix perdix)* with GnRH and na increase in the photoperiod. Bollettino delia Società italiana di biologia sperimentale, v. 60, n. 6, p. 1153, 1984.

MITCHELL, J. W. Heat transfer from spheres and other animal forms. Biophysical Journal, v. 16, n. 6, p. 561-569, 1976.

MYERS N, El atlas Gaia de la gestión del planeta. Madrid: Hermann Blume, 1987.

MYERS, S. A.; MILLAM, J. R.; EL HALAWANI, M. E. Plasma LH and prolactin levels during the reproductive cycle of the cockatiel *(Nymphicus hollandicus).* General and Comparative Endocrinology, n. 73, p.85-91,1989.

MOURA, DJ. Ambience in poultry production. In: SILVA, LJ.O. Ambiência na produção de aves em clima tropical. Jaboticabal: SBEA, 2001, v.2, p.75-148

NÃÃS, LA. Principles of thermal comfort in animal production. São Paulo, Ícone Editora, 183 p. 1989.

NÃÃS, I. A. Physical aspects of construction in the thermal control of the facility environment. APINCO Conference, Poultry Science and Technology. Santos: FACTA, p. 158- 167.1994.

NATIONAL WEATHER SERVICE CENTRAL REGION. Livestock hot weather stress. Regional Operations Manual Letter, p.31-76.1976.

NEUMANN, D.; KALETA, E. F.; LIERZ, M. Semen collection and artificial inseminationin cockatiels *(Nymphicus hollandicus)-/* potential model for psittacines. Tierãrztliche Praxis Kleintiere, v. 41, n. 2, p. 101-105, 2013.

NICKEL, R.; SCHUMMER, A.; SEIFERLE, E. Anatomy of the domestic birds. Berlin: Verlag Paul Parey, p. 95-97, 1977.

NOL, E.; CHENG, K.; NICHOLS, C. Heritability and phenotypiccorrelationsof behaviourand dominance rank of Japanese quail.Animal behaviour, v. 52, n. 4, p. 813-820,1996.

PEARSON, James T. Development of thermoregulation and posthatching growth in the altricial cockatiel *Nymphicus hollandicus*. Physiological and Biochemical Zoology, v. 71, n. 2, p. 237-244,1998.

PEREIRA R. J. G. Endocrine and behavioural monitoring of the annual reproductive activity of free-living male kiri falcons *(Falco sparverius)*. 83p. 2008. PhD thesis - Universidade Estadual Paulista, Faculdade de Ciências Agrárias e Veterinárias, Jaboticabal. 2008.

PERENCIN, F.; CUNHA, L.L; RIGOLETO, L.; MARTELLI, L.; COZEU, L.; BONICI, M.; GOMES, M. D.; MARTINS, T.; COSTA, T. A.; FAUSTO, T.; TAIRA, R. Information manual on responsible ownership of psittaciformes. 18p. 2011. Available at :

<http://www.fmvz.unesp.br/Home/UnidadesAuxiliares/CEMPAS/manual_psitacideos.pd f>. Accessed on: 01 June 2015.

PETERSON, R. T. The birds. Rio de Janeiro: J. Olympio, p. 208.1971.

PICKARD, A. R. Reproductive and welfare monitoring for the management of ex situ populations. Conservation Biology Series-Cambridge, p. 132-146, 2003.

POLLOCK, C. G.; OROSZ, S. E. Avian reproductive anatomy, physiology and endocrinology. Veterinary Clinics of North America: Exotic Animal Practice, v. 5, n. 3, p. 441-474, 2002.

POUGH, F. H.; HEISER, J. B.; McFARLAND, W. N. A vida dos vertebrados. São Paulo: Atheneu Editora, p. 439-448, 1999.

PRESTES, P. N. Description and analysis of the ethogram of Amazona pretrei in captivity. Ararajuba, v. 8, p. 25-42. 2000.

RITCHIE, B. W.; HARRISON, G.J.; HARRISON, L. R. Avian medicine: principies and applications. Lake Worth (FL): Wingers Publishing; p. 748-804,1994.

ROBBE, D.; TODISCO, G.; GIAMMARINO, A.; PENNELLI, M.; MANERA, M. Use of a synthetic GnRH analogue to induce reproductive activity in canaries *(Serinus canaria)*. Journal of Avian Medicine and Surgery, v. 22, n. 2, p. 123-126, 2008.

ROCHA, D. C. C. Behavioural characteristics of emus in captivity submitted to different photoperiods and different male: female ratios. 392f. 2OO8.PhD thesis. Federal University of Viçosa, 2008.

ROMAGNANO, A. Reproduction and paediatrics. In: BSAVA Manual of Psittacine Birds, 2nd Edition, Gloucester, 204p. 2005.

RUPLEY, A. E. Manual of avian practice. Philadelphia: WB Saunders, 1999.

RUTZ, F. Physiological aspects that regulate thermal comfort in poultry. Apinco Conference on Poultry Science and Technology. Campinas: APINCO Foundation for Poultry Science and Technology, p. 99-110, 1994.

SAAD, C. E. P.; FERREIRA, W. M.; BORGES, F. M. O.; LARA, L. B. Evaluation of expenditure and voluntary consumption of balanced rations and sunflower seed for true parrots (Amazona aestiva). Ciência e Agrotecnologia, v. 31, n. 04, p. 1176- 1183, 2007.

SAITO, N.; SHIMADA, K.; NOMURA, N.; OGURI, K.; SATO, K.jWADA, M.; Seasonal changes in the reproductive functions of Java Sparrows (Padda oryzivora). Comparative Biochemistry and Physiology Part A: Physiology, v. 101, n. 3, p. 459-463,1992.

SANDELL, M. I.; SMITH, H. G. Female aggression in the European starling during the breeding season. Animal Behaviour, v. 53, n. 1, p. 13-23,1997.

SANTOS, A. L. Q.; DE SOUZA, R. R.; MENEZES, L. T.; FERREIRA, C. H.; DE OLIVEIRA, S. R. P.; KAMINISHI, Á.; NASCIMENTO, L. R. Comparative anatomy of the digestive tract of different birds of the Psittaciformes order. PUBVET, v. 6, n. 13, 2012.

SAUVEUR, B. Photopériodisme et reproduction des oiseaux domestiques femelles. Productions animales-paris-institut national de la recherche agronomique-, v. 9, p. 25-34, 1996. SE

SCHEUERLEIN, A.; GWINNER, E. Is food availability a circannual zeitgeber in tropical birds? A field experiment on stonechats in tropical Africa. Journal of Biological Rhythms, v. 17, n. 2, p. 171-180, 2002.

SCHMIDT, R. E.; REAVILL, D. R.; PHALEN, D. N. Pathology of pet and aviary Birds. AMES: BLACKWELL PUBLISHING, p. 67-93. 2003.

SCHWARZE, E. Compendium of veterinary anatomy - anatomy of birds. ZARAGOZA: ACRIBIA, 212p. 1980.

SCHWARZE, E.; SCHRÕDER, L. Compendium of veterinary anatomy. Zaragoza : Acríbia, 1970. V.5, p.145

SEIBERT, L.M. 2003. "Social dominance: The peck order revealed." Proc Assoe Avian Vet, Pittsburgh, PA, p. 187-188, 2003.

SEIBERT, L. M.; CROWELL-DAVIS, S.L. Gender effects on aggression, dominance rank, and affiliative behaviours in a flock of captive adult cockatiels (Nymphicus hollandicus). Applied animal behaviour Science, v. 71, n. 2, p. 155-170, 2001.

SHARP, P.J.; DAWSON, A.; LEA, R. W. Control of luteinising hormone and prolactin secretion in birds. Comparative Biochemistry and Physiology Part C: Pharmacology, Toxicology and Endocrinology, v. 119, n. 3,

p. 275-282,1998.

SHARP, P. J.; SREEKUMAR, K. P. Photoperiodic control of prolactin secretion. Avian endocrinology, p. 245-255, 2001.

SHIELDS, K. M.; YAMAMOTO, J. T.; MILLAM, J. R. Reproductive behaviour and LH levels of cockatiels (Nymphicus hollandicus) associated with photostimulation, nest-box presentation, and degree of mate access.Hormones and behaviour, v. 23, n. 1, p. 68-82, 1989.

SICK, H. Ornitologia brasileira. Rio de Janeiro: Nova Fronteira, p. 113-360, 1997.

SILVA, R.G. Introduction to animal bioclimatology. São Paulo: Nobel, 286p. 2000.

SILVA, M. A. N.; HELLMEISTER FILHO, P.; ROSÁRIO, M. F.; COELHO, A. A. D.; SAVINO, V. J. M., GARCIA, A. A. F., SILVA, I. J. O.; MENTEN, J. F. M. Influence of rearing system on performance, physiological condition and behaviour of broiler strains. Revista Brasileira de Zootecnia, Viçosa, v. 32, n. I, p. 208-213, jan./feb. 2003.

SILVA, I.J.O.; NÃÃS, LA. Influence of afforestation on the thermal performance of poultry houses through thermal comfort indices and egg production. In: CONGRESSO BRASILEIRO DE ENGENHARIA AGRÍCOLA, Poços de Caídas, p.231-233.1998.

SMITH, G. A. Neptune City. In: The encyclopedia of cockatiels. New Jersey: TFH, 1978. SMITH, T.R.; CHAPA, A.; WILLARD, S.; HERNDON JR, C.; WILLIAMS, R.J.; CROUCH, J.; RILEY, T.; POGUE, D. Evaporative Tunnel Cooling of Dairy Cows in the Southeast. II: Impact on Lactation Performance. Journal of Dairy Science, v.89, n.10, p.3915-3923, 2006.

SOUZA, J. L. M. Manual de Construções rurais. Curitiba : DETR/SCA/UFPR, 165 p. 1997.

SOUZA, P. Avicultura e clima quente: como administrar o bem-estar aos aves? Avicultura Industrial, Porto Feliz, n.4, p.52-58, 2005.

STALEY, A. M.; BUKNCO, J. M.; DUFTY, A. M.; WILDT, D. E.; MONFORT, S. L. Faecal steroid monitoring for assessing gonadal and adrenal activity in the golden eagle and peregrine falcon. Journal of Comparative Physiology B: Biochemical, Systemic, and Environmental Physiology, v. 177, n. 6, p. 609-622, 2007.

STERLING, R.; SHARP, P. A comparison of the luteinising hormone-releasing activities of synthetic chicken luteinizing hormone-releasing hormone (LH-RH), synthetic porcine LH- RH, and Buserelin, and LH-RH analogue, in the domestic fowl. General and comparative endocrinology, v. 55, n. 3, p. 463-471, 1984.

SWEENEY, R. G. Husbandry, breeding and hand-rearing of Salmon-crested cockatoo at Loro Parque Fundacion, Puerto de la Cruz. International Zoo Yearbook, v. 37, n. 1, p. 130- 137, 2000.

TAKAHASHI, L. S.; BILLER, J. D.; TAKAHASHI, K. M. Bioclimatologia Zootécnica. Unesp, Jaboticabal, 91p. 2009.

TAO, X.; XIN, H. Acute synergistic effects of air temperature, humidity, and velocity on homeostasis of market-size broilers. Transactions of the ASAE, St. Joseph, v.46, n.2, p.491-7, 2003.

TARVIN, K.A.; WOOLFENDEN, G.E. Patterns of dominance and aggressive behaviour in blue jays at a feeder.

Condor, p. 434-444,1997.

TCPO 7. Price composition tables for budgets.Editora Pini, São Paulo, p.830,1980.

TINÔCO, I.F.F. Avicultura industrial: novas conceitos de materiais, concepções e técnicas construtivas disponíveis para galpícolas brasileiros. Revista brasileira de ciência Avícola, v. 3, n. 1, p. 01-26, 2001.

THOM, Earl C. The discomfort index. Weatherwise, v. 12, n. 2, p. 57-61,1959.

TORLONI, C. E. E. Breeding cockatiels. São Paulo: LIS Gráfica e Editora, 1991. 89 p.

TSCHUSDIN, A.; RETTMER, H.; WATON, H.; CLAUSS, M.; HAMMER, S. Evaluation of hand- rearing records for Spix's macaw Cyanopsitta spixii at the Al Wabra Wildlife Preservation from 2005 to 2007.International Zoo Yearbook, v. 44, n. 1, p. 201-211, 2010.

UBUKA, T., BENTLEY, G. E. Neuroendocrine control of reproduction in birds. Hormones and Reproduction of Vertebrates: Birds, p. 1-25, 2011.

VERDADE, L. M. The exploitation of wild fauna in Brazil: alligators, systems and human resources. Biota Neotropica (Ed. Portuguesa), São Paulo, v. 4, n. 2, p. 5-17, 2004.

VIEIRA, F. R. M.; COSTA, F. M.; TASSI, V. M.; BOLOCHIO, C. E.; CUNHA, I. P.; ASSATO, E. H.; SOUZA, C. A. I.; MAGALHÃES, F. C.; MACHADO, C. S.; CELEGHIN, P. C.. Manual for keepers -   Guarulhos   Zoo.   38p.
          2008. Available at:
<http://szb.org.br/blog/conteudos/bibliografias/07-manejo/manual-para-tratadores-   zoo-guarulhos.pdf>. Accessed on: 07 June 2015.

VIGODERIS, R. B. Development of a prototype for adiabatic evaporative cooling in air-conditioned animal facilities using expanded clay. 53 p. 2002. Master's dissertation - Federal University of Viçosa. 2002.

VLECK, C. M. Hormones, reproduction, and behaviour in birds of the Sonoran desert. Avian endocrinology, p. 73-86,1993.

YAMAMOTO, J.T.; SHIELDS, K.M.; MILLAM, J.R.; ROUDYBUSH, T.E.; GRAU, C.R. Reproductive activity of force-paired cockatiels (Nymphicus hollandicus). The Auk, p. 86- 93,1989.

YANAGI JÚNIOR, T. Technological innovations in animal bioclimatology aimed at increasing animal production: the relationship between animal welfare and climate. 2006. Available at: <http://www.infobibos.com/Artigos/2006_2/ITBA/Index.htm>. Accessed on: 05 June 2015. YANAGI JÚNIOR, T.; DAMASCENO, G. S.; TEIXEIRA, V. H.; XIN, H. Prediction of black globe humidity index in poultry buildings. VI Livestock Environment- International Symposium: ASAE, Louisville, KY, p. 482-489, 2001.

WASSER, S.K.; HUNT, K. E. Noninvasive measures of reproductive function and disturbance in the barred owl, great horned owl, and northern spotted owl. Annals of the New York Academy of Sciences, v. 1046, n. 1, p. 109-137, 2005.

WILSON, E. O. Sociobiology - the new synthesis. Cambridge, MA: Harvard University Press, 697p. 1975.

WILSON, J.D. Correlates of agonistic display by great tits Parus major.Behaviour, v. 121, n. 3, p. 168-214,1992.

WINGFIELD, J.C.; BALL, G.F.; DUFTY, A.M.; HEGNER, R.E.; Ramenofsky, M. Testosterone and aggression in birds. American Scientist, n. 75, p.602-608, 1987.

WOOLFENDEN, G.E.; FITZPATRICK, J. W. Dominance in the Florida scrub jay. Condor, v. 79, p. 1-12, 1977.

ZANN, R. Behavioral studies of the quarrion *(Nymphicus hollandicus).* University of New England, Armidale, 1965.

I want morebooks!

Buy your books fast and straightforward online - at one of world's fastest growing online book stores! Environmentally sound due to Print-on-Demand technologies.

Buy your books online at
www.morebooks.shop

Kaufen Sie Ihre Bücher schnell und unkompliziert online – auf einer der am schnellsten wachsenden Buchhandelsplattformen weltweit! Dank Print-On-Demand umwelt- und ressourcenschonend produziert.

Bücher schneller online kaufen
www.morebooks.shop

info@omniscriptum.com
www.omniscriptum.com

Printed by Books on Demand GmbH, Norderstedt / Germany